Nifty E-Z Adventures in DXing

Developing Worldwide Friendships through Radio Communication

First Edition

By Jeffrey A. Cantor, K1ZN

Another guide in the
Nifty! Ham Accessories
Easy Guide Series

www.niftyaccessories.com

Copyright

Copyright © 2015 Dr. Jeffrey A. Cantor, K1ZN. All rights reserved. No part of this book or portions thereof may be reproduced in any form or by any means, electronic or mechanical, including photocopying, recording, or by any other means, without permission in writing from the publisher.

Published by

Nifty Ham Accessories, 2015
www.niftyaccessories.com

Disclaimer and Limitation of Liability

While every effort has been made to make this publication as accurate as possible, Nifty! Ham Accessories and the author assume no liability for the contents regarding safety or damage to equipment or person, and do not guarantee the accuracy herein.

Icom, Elecraft, Ten-Tec and Buddipole are registered trademarks of these corporations.

Contents

Special Thanks To 6

Chapter 1: What's DX All About 9
   The Lure of Long Distance Communication 9
   Working the World 10
   Countries and Entities 10
   Paper Chasing and Making Friends 11
   Contesting 13
   Vacation Operating 13
   Becoming a DXer 14

Chapter 2: DX Operating Practices 17
   Effective Listening 17
   Know Your Radio 17
   Where's the DX? 18
   DX Spotting Clusters 19
   Finding the DX Listening Frequency when Working Split 20
   Working Pileups & Splits 21
   Calling DX Stations 22
   Know Your Phonetics 22
   Know Your Prosigns 23
   Frequency Interference - QRM, UQRM and DQRM 23
   DX Code of Conduct 24
   Lessons from DXpeditions – Navassa 27

Chapter 3: Station Requirements 29
   Transceivers and Options 29
   Transceiver Selection Checklist 30
   Additional Operating Accessories 30
   Antennas and Towers 32
   Lightning Protection 34
   QRO / High Power (or not) 35
   Amateur Radio World Map & Reference Materials 35
   Know the CQ & ITU Zones 36
   Using Contests to Catch DX 38

DX Nets and Call Lists..41
Remote Operating 43

Chapter 4: Propagation 45
Propagation Basics 45
Propagation Varies by Band..47
What do Propagation Forecast Numbers Mean?.....................49
Working Greyline...51
Long Path 51
Skew Path 52
Sporadic-E Propagation...52
Propagation Prediction Resources...52
Beacons 52
Reverse Beacon Network 53
VOACAP, HF Propagation Prediction Software 54

Chapter 5: Spotting Clusters and DX Publications 55
Online Spotting Programs & Operating Resources.......................55
DX Clusters - Use & Etiquette...57
Additional Online Resources 60

Chapter 6: Documenting the Contact 65
Keeping a Log 65
QSL Logging and Contesting Software 66
Online QSL Lookup Databases...68
QSL Managers..69
ARRL Outgoing & Incoming QSL Bureaus 69
QSL Direct with SASE 70
OQRS & ClubLog...71
LogBook of the World (LOTW)...73
E-QSLs 74
DXCC & IOTA Card Checking 74

Chapter 7: Awards and Paper Chasing 75
DXCC Award..75
Worked all States Award...77
VHF/UHF Awards 78
Islands on the Air Program & Award...79
Worked all Continents Award (WAC)..81

Worked All Zones Awards..82
CQ Worked All Zones, WAZ 82
Various Countries Also Sponsor Awards...84
Stateside Operating Awards 84

Chapter 8: DXing on the Top and Magic Bands....................89
Top Band DXing 89
160M Propagation Characteristics 89
Listening Effectively 89
Listening Antennas 90
Transmit Antennas 91
On 160 QRO is a Must..91
K1ZM's Recommendations 92
Six Meters - the Other End of the Spectrum 92
Six Meter Propagation Characteristics 93
Equipment Needs 93
Awards & Programs 93

Chapter 9: Unique Challenges of QRP 95
Integrating QRP into Other Activities..95
QRP Transceivers – Factory Made or Build Your Own 95
Portable Antennas 97
Signing QRP...98
QRP Organizations & Groups...98
Where to find QRP Stations 99
Operating Activities 99
Awards & Certificates – QRP 100
Tips for QRP DXing or Contesting...100

Chapter 10: Considerations for Going QRO 101
Increased Power Operation 101
Linear Amplifiers 102
Other QRO Station Considerations 104

Chapter 11: DX Clubs and Organizations 105
Joining a DX Club...105
Fundraising for Dxpeditions..106
Major DX Get-togethers...107
Buy/Sell/Swap Resources 108

Appendix A: **Entities of the World**	110
Footnote Citations	117

About the Nifty E-Z Guide to Adventures in DXing

Using easy to understand language and illustrations, this guide describes how to outfit your station with the goal of making worldwide DX contacts and applying for various DXing awards.

With the vagaries of constantly changing radio wave propagation characteristics, coupled with seasonal and hourly variations, chasing DX can be an endlessly fascinating endeavor.

It can be a great source of pleasure in being able to use one's own skills and equipment to make contact with some far-off land, establishing lasting worldwide friendships.

As one progresses in the art of DXing and becomes successful in making worldwide contacts, it's a natural progression to want to work towards fulfilling the requirements for applying for various awards, receiving personalized certificates and plaques recognizing your accomplishments.

This book has been written expressly to help you enjoy the DXing hobby by successfully outfitting your station, completing contacts and applying for awards.

Let's get started!

Special Thanks To

First and foremost, to my YL, my wife Ruth, who has supported me in this hobby for almost 50 years. She has allowed family decisions about where to live be influenced by the ability to erect tall towers, wire antennas and the like. Ruth I love you with all my heart!

To my Southeastern DX Club members for their advice and guidance, and sharing of their unique experiences. To my friends Hal Kennedy, N4GG and Jack Ray, W4JBR, who both reviewed the chapter on propagation. In addition, Hal also reviewed and edited the first three chapters for me. Thanks, guys.

Jeff Briggs, K1ZM, the "top band guy" read and made suggestions about chapter 8 on 160M DXing. Jeff, I thank you for your longstanding support and guidance.

I thank my friend Barrett Thompson, KE4R, a true sportsman and QRP operator for his input on the QRP portion of Chapter 9.

John Tramontanis, N4TOL, my fellow SEDXC club member who graciously agreed to review chapters 1, 2, 3 and 5, 8 and 9. Thank you very much, John.

The Southeastern DX Club's president, Wey Walker, K8EAB, for reviewing Chapters 5, 6 and 7.

And I also want to thank those organizations and companies who have extended permission to use their information, illustrations or product representations in this work. These include Eric Swartz, WA6HHQ of Elecraft Corp., Chris Drummond of Buddipole Company, Glen Johnson, W0GJ and the Grey Line Report, Rich Moseson, W2VU of CQ Magazine, James Thompson of Radioworks, Inc., Paul Herrman, N0NBH, Jeff Briggs, K1ZM, Tim Makins, EI8IC, Ten-Tec Corp., Icom Corp., Michael, G7VJR and ClubLog, DX Summit, OK7K Contest Team, U.S. Original 13 Colonies Group, Dan Sullivan, W4DKS and the RSGB & the IOTA program, Bruce Horn, WA7BNM, DX Coffee, Dan Sullivan, W4DKS, David Solomon, AG4F, Press Jones, N8UG, Giles Read, G1MFG, Frank

Beafore, WS8B, Dennis J. Lusis, W1LJ/DL, Edwin Jones, AE4TM, Bryce K. Anderson, K7UA & his "New DXer's Handbook," Ian Poole, G3YWX, James Kearman III, KR1S, Mark Downing, WM7D, Hamuniverse.com, Thomas Schrage, W0ZV, Steve San Andrea, AG1YK, Northern California DX Foundation, International DX Association, YASME Foundation, Scott Neader, KA9FOX, NCDXA/IARU, & ARRL.

And last, but not least to Bernie Lafreniere, N6FN, of Nifty! Ham Accessories for accepting my manuscript for publication, editing the work and formatting it for printing.

Chapter 1: **What's DX All About**

Ham radio has been my hobby for almost 45 years and I enjoy interacting with other hams around the world. My interest in long range communications began as a shortwave radio listener. Some of you might remember "Radio Moscow," and BBC broadcasts on the "short waves." As a young person I was intrigued by different perspectives on issues and topics often discussed on these stations. I also recall listening to hams from far-away places, instilling in me a desire to be become a ham radio operator myself.

The Lure of Long Distance Communication

The quest to seek out radio contacts with different and diverse countries, islands, and other geographic and political "entities" became a source of great pleasure and challenge for me – and for many of my close friends. This aspect of the Amateur Radio hobby is commonly referred to as "chasing DX." My vacations and world travels are often planned in pursuit of this hobby.

Amateur Radio communication promotes friendship and good will among its participants. As we make international radio contacts, we learn more about each other as persons, and acknowledge that we are more alike than different. Thus, through worldwide Amateur Radio communications we promote international good will, friendship and understanding. Many DXpeditions are humanitarian in nature; they are planned with the intent of bringing health care, food, and other life-saving services to distant locations, in addition to playing "radio." For example, the 2015 E30FB DXpedition supported Eritrea's participation in the 2020 Tokyo Olympic and Paralympic Games.

Enjoying the challenge of DX is what this book is all about. As part of an effort to share this enthusiasm for long range communication, I support Amateur Radio as a volunteer license examiner (VE). We see numbers of people come into the hobby to get licensed to use 2M handheld radios, but too few of them venture beyond local repeaters and onto the HF bands. I chose to write this book to introduce and encourage fellow ham radio hobbyists into chasing DX and to help them succeed in the challenge, hopefully becoming an enthusiast of this aspect of Amateur Radio.

Working the World

Several hundred countries including island nations, plus political subdivisions and religious entities provide a ready supply of destinations to attempt to contact. When I began chasing DX one of my goals was to see how far I could transmit signals to distant lands. I would experiment with wire antenna configurations that I read or learned about, always attempting to improve communications capability.

As I studied world history and geography in school, the aspect of actually making contact with peoples in these far off places reinforced my studies and motivated my learning. A useful benefit to this aspect of ham radio is learning more about our world. As I worked a country (England, Canada, Egypt) I found it fascinating to research interesting facts about the place, and of course locate the country and the specific station on a map. Exchanging QSL cards with the other station would prompt me to research additional details of the new location. QSLing practices are explained later in the book.

In retirement, my wife and I enjoy travelling to places and visiting friends made on the air. A recent trip to Zurich was significantly enriched by spending some time with a friend made on the radio. It's surprising how much fun can be had when a local helps you enjoy the things to see and do in their country, perhaps even inviting you to operate from their home station.

Countries and Entities

Today, the official number of these world countries or places we call entities (the ARRL DXCC term) is 340.[i] Additionally, over time there were 61 other entities that "came and went" based on political or geographical changes, etc.

I also become familiar with political entities such as the Vatican, which is a separate DXCC entity, physically located inside of Italy. Other entities include United Nations Headquarters (in New York City), UN Geneva (inside Switzerland), Sovereign Military Order of Malta (a religious order located in Italy), and Mount Athos (a religious order inside Greece).

Personalized QSL Card

Some island nations that exist as entities include: Hawaii, Puerto Rico, Jersey, Wales, Sicily, etc. Minami Torishima is an island belonging to Japan that I had been chasing for several years and finally managed to work as I was writing this chapter. Distant lands, and sometimes not-so-distant lands and entities provide abundant opportunities for making radio contacts and developing new friends, perhaps becoming regular on-the-air buddies.

Paper Chasing and Making Friends

One of the goals of a DXer, is to make two way radio contact with each of the official 340 entities. A variety of certificates and plaques can be applied for as you work towards that goal. The most recognized award program and forum for chasing DX is called the DXCC Program sponsored by the American Radio Relay League. The first benchmark award is the DXCC Century Club certificate, - earned by verifying contact with 100 of the currently available 340 entities.

DXCC Award
Courtesy of David Salomon, AG4F

Once I attained DXCC with 100 verified two way contacts (QSL cards in hand) across the HF bands (80-10M), and mixed modes of operation (CW, Phone, and/or Digital), the next benchmark was to attain 100 contacts on each of those bands. That award is the DXCC Century Club 5-Band award. Of course, as I was aspiring to 5-band DXCC, I was also working towards endorsements such as 100 on each of the WARC bands (17, 12 & 30M) and DXCC–CW and DXCC-Digital. Additionally, while chasing these awards a DXer continues to amass additional entities towards the next major milestone of DXCC Honor Roll – 331 of the 340 current entities, nine less than the total number of entities.

Island nations and entities provide additional paper chasing opportunities. I am particularly excited about the award programs sponsored by the Radio Society of Great Britain called the *Islands on the Air Program* or IOTA.

The IOTA program recognizes some 1200 worldwide island groups. Each of these is assigned a unique identifier called an IOTA number. For example, the Island entity of Puerto Rico is NA-099, the "NA" being the designation for North America. This award program's basic award structure is in certificates per 100 island groups, plus certificates per continent.[ii]

IOTA QSL, Marabatuan Island

Contesting

Between chasing *all-time-new-ones* and other DX, many DXers also participate in on-the-air radio contests of multiple varieties, from counties by state, to international radio callsign prefixes. Besides being fun, contesting is a way of honing your operating skills. Contesting and its relationship to chasing DX are discussed more fully in Chapter 2.

Vacation Operating

Ever thought about going to an island or maybe a semi-rare country and taking along your radio transceiver? More and more hams are doing this and enjoying the experience. If this appeals to you, pick a location that interests you. Then visit the ARRL web page where you will be able to determine if a reciprocal licensing agreement is in place with that country or entity that allows you to operate from that location with your callsign, perhaps with a prefix preceding your call.

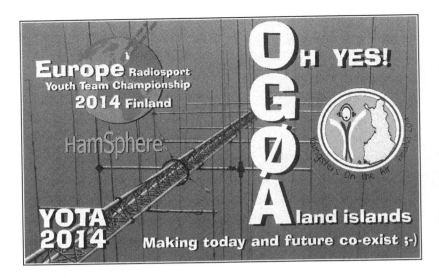

Becoming a DXer

The information, practices and hints contained in this book will make your efforts at chasing DX both fun and productive. Beginning with DX on the air operating practices: the art of listening well, locating and identifying DX signals, dealing with on background noise and interference, understanding the DX practice of split-frequency operation, and finally successful calling tactics for working the station will be discussed.

As hams and DXers, our station equipment changes over time. As we gain operating experience we tend to acquire more sophisticated transceivers that enable us to enhance operating proficiency and pull in the weak ones. We also experiment with building or buying better antennas. I will share some insights gained through personal experience and by association with some of the hobbies' "big guns." A newly emerging tool for DXers is remote operating, and some insights learned about using remotely located transceivers will also be provided.

The pursuit of DX promotes the art and science of radio. All DXers experiment with antennas, and many with radio equipment modifications. We learn from each other in furtherance of the hobby On some DXpeditions, operators visit local schools or invite loca school children to the site of the DXpedition to learn about ham radic and in some cases provide local license instruction, and in this way both promote Amateur Radio and expose young people to the world of international radio communication.

An understanding of radio wave propagation is essential for efficiently working DX and some insights and tips are provided later in this book. Sources of propagation data and the use of prediction software will also be highlighted.

I will also discuss using nets, lists and publications as a means of growing your numbers of worked entities. Various Internet publications are available that can inform us of upcoming DXpeditions, which are a great way of adding to your list of contacted entities.

Chapter 2: **DX Operating Practices**

You can't work them if you can't hear them! This is the basic principle of DXing. All the power in the world will not compensate for not being able to hear a station. Therefore, as I point out later in the book, antennas are important considerations for a good DX chasing station. However, techniques for effective listening on the band and frequency are equally important.

Effective Listening

The high frequency bands can be noisy. Noise comes from natural circumstances, the atmosphere, weather, etc. Noise can also be caused by other stations on or nearby your listening frequency. The low bands (160M - 80M – 40M) are more susceptible to natural noise than the high frequency bands (20M – 15M – 17M – 10M). How you tune in and listen on a given frequency will determine your ability to know that a station is there and that you have made contact.

Your concentration on the tuned signal is paramount. I found by experience that using a set of noise-cancelling headphones to block out background noise, helps focus ones listening on a single signal. With practice you can train your brain to filter out extraneous adjacent signals and noise.

Know Your Radio

Learn to use the bandwidth and noise cancelling digital signal processing (DSP) capabilities of your transceiver. Practice using these filters and DSP settings in tandem on difficult to hear signals, the net effect can be quite effective at bringing in an otherwise unreadable signal. Just like driving a car and knowing its controls and features, you need to know your radio's controls and built in filtering capability. Try tuning in a weak signal during noisy band conditions. Try backing off the RF gain while increasing the AF gain to improve readability of the signal.

The operator's manual that came with the rig should describe how to use the RF, AF, and DSP controls, as well as the various filter combinations. The series of radio operating guides from Nifty Ham Accessories do a good job of explaining these controls and how they interact. You will be well served by getting one of their Mini-manuals for your model radio. You should be able to find one for your radio at most ham radio stores and on the manufacturer's web page www.niftyaccessories.com

Fine tuning a signal can be accomplished through the RIT/XIT controls which permits tuning either/or both receive and transmit frequencies in very small increments. If the signal you are receiving drifts off the original contact frequency, do not tune your VFO, as your transmit frequency will change as well. The solution is to use RIT, which will fine tune your receive frequency without altering the transmit frequency. This can be especially useful when you are the one calling CQ and the responding station comes back off frequency. To prevent altering your transmit frequency, use your RIT control to tune the station in.

To maximize signal readability, good noise-cancelling headphones are essential; you cannot effectively focus on hard to hear DX signals using the transceiver's external speakers. It's very difficult to concentrate on hearing weak signals while being distracted by background noises such as fans, relay clicks, TV's, street noise and the like. Ask your fellow ham friends what brand and model of headphones or headsets (earphones and mic) they use, and what they recommend.

There are some very well engineered products on the market today, with embellishments that maximize audio quality. Either visit your local ham radio store or the booths of manufacturers at a hamfest to look at and try different headphones and headsets. In addition to audio quality and noise reduction, pay particular attention as to how well they fit and if they will be comfortable to wear for a long period of time. A headset is also essential for operating VOX, which is useful in being able to both transmit and listen, leaving your hands free to enter QSOs into your log instead of having to operate a push-to-talk (PTT) mic.

Where's the DX?

Perhaps the easiest way for a ham to make a radio contact is to get onto a vacant frequency and call CQ. In practice, this means ensuring that the frequency that you are calling on is not in use by first listening and then transmitting "is this frequency in use, this is K1ZN?" Then, if hearing no reply, transmit "CQ DX, CQ DX, CQ DX, this is K1ZN, K1ZN, K1ZN." You will be pleasantly surprised to discover that you can work a fair number of DX entities early on in your quest for DXCC in this way.

Additionally, listening on the bands for overseas or distant stations will also work. When I first began DXing, the practice was to tune across the bands and listen for distant stations and foreign calls. This still works today. From time to time I will simply select a band based upon the time of day and propagation conditions, starting at one end of the band, either the CW portion or phone portion, and listening carefully while tuning across the band segment.

Yes, listen. Listen for stations that might be calling for anyone to respond (CQ) or for a specific kind of station (CQ NA, or Japan only, etc.). Sometimes they will call for stations by the numbers (the number in the callsign). Particularly listen for the DX station's callsign. If it is a foreign call prefix – such as HB (Switzerland), check your log to see if you might need it for an ATNO – *all-time-new-one*? Or as a band fill on that particular band?

Listen also for stations that might be working stations very rapidly – one after another with only signal reports and thank-yous. This is typical of DX activity. Listen for that station's call – is it a DX station? Are they working split (transmitting on one frequency and listening on another), which is also typical when DX stations are managing a pile-up? Can you hear responding stations, or not? And, do you need that station? A quick note here, if the DX is working split, never call them on their transmit frequency. Either wait for them to announce where they are listening (up 3 or whatever) or tune up and down from their transmit signal a bit to find where stations are calling the DX station.

DX Spotting Clusters

Packet clusters began appearing across the DX community as the age of computers emerged. Packet clusters originally were electronic bulletin boards primarily using VHF frequencies to transmit their data. These clusters eventually morphed into the worldwide web networks of today, which are very easy to use.

I monitor the *"DXSummit"* cluster **www.dxsummit.fi**, one of the better known clusters. This DX spotting page is very versatile and allows me to monitor all worldwide spots, or limit the spotting by: band, DX entity, mode, etc. e-Ham.net supports another popular spotting service.

With tools like one of these, it is easy to identify desired DX stations, tune to the frequency indicated and work that station.

Later in the chapter I discuss working DX on DX nets, or scheduled and moderated groups designed to facilitate working DX. There are some DXers who frown on this process – but there is nothing wrong with the process and it merits discussion.

Finding the DX Listening Frequency when Working Split

The more rare and sought after DX will often transmit on one frequency and listen elsewhere in the band. Thus, you need to master the process of operating "split." What's that? Working split frequency is simply listening to the incoming signal from the DX station on a given frequency, and responding (transmitting) back to that station on an adjacent frequency as the station directs or specifies.

For example, you may spot a DX station operating CW on 14.007 MHz who calls QRZ NA (North America) listening up 3. That means he/she is listening for stations calling on 14.010 MHz, 3 kHz up from their transmit frequency. You would then activate your transceiver's second or "B" VFO for transmit and tune it to 14.010 MHz. Then use VFO "B," to transmit to the DX station upon his/her request for calls while listening for their response and acknowledgement of your call on VFO "A." Instructions for doing this vary depending upon the model of radio you have, either refer to the manufacturer's manual or the radio guide from Nifty Accessories, **www.niftyaccessories.com**

In most sought after DX stations, the split frequency is a span of frequencies; usually up from the DX station's transmit frequency. In CW mode it might range from 1 – 3 or more kHz, in SSB its usually 5 – 10 kHz or more. Digital mode could be somewhere in between.

Very rare DX will often have more extensive split spectrum. This is purposely done to spread out the calling stations and make it easier for the DX operators to hear and respond to calling stations. In the case of 40M, it is not uncommon to find the DX station transmitting phone on 7.085 MHz which is not a phone frequency in the U.S., but listening for U.S. stations on 7.185 MHz which is a phone frequency in the U.S. Chapter 4 discusses what bands to be on for DX during what times of day or evening.

Working Pileups & Splits

Pileups happen when many DXers are calling the DX station simultaneously. That is why DX stations use a split frequency process. Where you position yourself in the split spectrum will ultimately determine your success to working the DX. If the DX station is listening across a spectrum of frequencies, how can I locate him/her? The answer is to adopt an operating strategy and execute same. One strategy might be to select a frequency point in the middle of the announced split (i.e. up 2 – 4 KHz, select up 3 KHz). Then respond to the DX station on that frequency, and continue responding on that frequency until he finds you. I used this scheme for many years, and very successfully at that.

However, with the advent of the spectrum scope or panadaptor, we now have a visual method of seeing adjacent spectrum signals as we operate. Therefore, we can visually see where the DX station is responding to stations, and strategically move our transmit frequency to a nearby spot. Panadaptors have proven to be extremely useful DX tools. Unfortunately only a few radios currently have this capability.

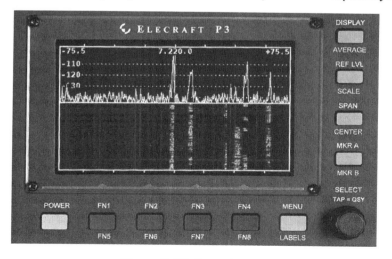

Elecraft P3 Panadapter
Courtesy of Elecraft Corporation

Calling DX Stations

What is the best way to call or respond to a DX operator to maximize your chances of getting his attention? What do you say? DXing is much like a sport. As in any sport there are certain conventions that come into play. In DXing it is the custom to exchange calls and signal reports only. Brevity is essential, as the DX and the many stations wanting to work the DX want to make as many contacts as possible while favorable band and propagation conditions exist. So, the process is to listen to the DX station, and transmit your call as directed (they may be calling only Europe, or only North America, or by callsign numbers, and/or they may be working split); listen for your call to come back.

Sometimes the DX station may only respond with a portion of a call that might be yours, -- "ZN station, station with ZN in the call;" you would then say your call again to determine if it was you they were calling. If the DX comes back to you, then respond with a signal report – which is usually 599 or 5NN (CW), regardless of actual band conditions. Don't give your name, QTH, or anything else, unless asked for.

Know Your Phonetics

Radio Amateurs are expected to know and use international radio phonetics, which are standardized as shown below.

	International Radio Phonetics				
A	Alpha	J	Juliet	S	Sierra
B	Bravo	K	Kilowatt	T	Tango
C	Charlie	L	Lima or London	U	Uniform
D	Delta	M	Mike	V	Victor
E	Echo	N	November	W	Whiskey
F	Foxtrot	O	Ontario or Oscar	X	X-ray
G	Gulf or Germany	P	Papa	Y	Yankee or Yokohoma
H	Hotel	Q	Quebec or Queen	Z	Zulu or Zanzibar
I	India	R	Romeo or Radio		

Know Your Prosigns

DXers have a language all of their own. Much of it is derived from Morse code shorthand – the sending of shortened words and special abbreviations to make our wrists less tired and get the message through in fewer words, etc. Some of it was borrowed from Western Union Telegraph operators, as well. The Q-Code evolved out of that practice and is in use both on phone and CW today. The table below summarizes the most commonly heard Q-Codes.

Common Q-Signals / Prosigns	
QRG = What is the exact frequency? The exact frequency is...	QRK = What is the readability of my signals? Readability is ...
QRM = There is interference on frequency, or on an adjacent frequency. Are you being interfered with?	QRN = There is static or noise on frequency. Is there static or noise on frequency?
QRO = Shall I increase power? I am increasing power. Today it also can mean, I am using a linear amplifier.	QRP = Shall I decrease power? I am decreasing power. Today, it can also mean, I am using 5W or less.
QRT = I am ceasing transmission. Shall I stop transmitting?	QRV = Are you ready to receive?
QRX = Wait, stand-by. Stop transmitting.	QRZ = Who is calling? Please call again.
QSB = Your signals are fading. Are my signals fading?	QSK = Can you hear me between your signals? I can hear you between my signals.
QSL = Message acknowledged. Can you acknowledge my transmission?	QSO = A two-way communication.
QSY = Please change frequencies. I am changing frequency.	QTH = Location of my shack. What is your shack location?

Frequency Interference - QRM, UQRM and DQRM

As DXers we are concerned with an increasing amount of on the air deliberate interference caused by people who gain some sort of satisfaction in making ham radio difficult for others. Man-made interference is termed QRM. Deliberate interference, which is a federal offense punishable by a hefty fine, is termed DQRM. On the air mistakes, or unintended interference, though still bad is termed UQRM.

The Amateur Radio community is attempting to formulate plans and strategies to identify and eliminate this interference – a huge task to do worldwide. The Navassa (K1N) DXpedition was the first to operationalize a preliminary process for DQRM identification. Be on the lookout for more about that project as it becomes available.

DX Code of Conduct

At this point in our discussion, I want to call your attention to the DX Code of Conduct that hopefully all of us practice. I trust this is not new information for you.

DX Code of Conduct

I will listen, and listen, and then listen again before calling.

I will only call if I can copy the DX station properly.

I will not trust the DX cluster and will be sure of the DX station's callsign before calling.

I will not interfere with the DX station nor anyone calling and will never tune up on the DX frequency or in the QSX slot (where the DX is listening).

I will wait for the DX station to end a contact before I call.

I will always send my full callsign.

I will call and then listen for a reasonable interval. I will not call continuously.

I will not transmit when the DX operator calls another callsign, not mine.

I will not transmit when the DX operator queries a callsign not like mine.

I will not transmit when the DX station requests geographic areas other than mine.

When the DX operator calls me, I will not repeat my callsign unless I think he has copied it incorrectly.

I will be thankful if and when I do make a contact.

I will respect my fellow hams and conduct myself so as to earn their respect.

We all want to engage in the sport of contacting distant radio stations. We want to be that "top gun" station that scores the contact first. Some of us want to be the loudest station on the air with the biggest beam antennas and most powerful amplifiers. Some of us can't afford to be that large and powerful and need to compete through good operating practices and prowess. In any event, we all need to subscribe to the DX Code of Conduct.

❖ **I will listen, and listen, and then listen again before calling.**

Seasoned radio amateurs will tell you that you cannot work a station if you can't hear it. You won't know who else is working that station if you don't listen carefully. You might not even realize that the station is listening on another frequency from that on which they are transmitting (working split). Therefore, it is essential for all involved that you learn to listen, listen and listen before hitting your transmit button.

❖ **I will only call if I can copy the DX station properly.**

There is no point in trying to call a DX station if you cannot hear that station well enough to know that the station is responding to your call. We are all surprised to learn that our signals often make it to far-off places, sometimes well beyond our own receiving capabilities. If you cannot copy a station well enough to be able to recognize your call coming back to you, stand by on frequency for a while and wait for a change in band conditions in your favor.

❖ **I will not trust the DX cluster and will be sure of the DX station's callsign before calling.**

Again, listen, listen, and listen again. Be sure you know who you are listening to. Just because you saw a spot of a DX station you need or want on a particular frequency, don't assume that is who you are hearing on that frequency. Verify the callsign of that station by listening on frequency, and then make your call when the station calls QRZ.

❖ **I will not interfere with the DX station nor anyone calling and will never tune up on the DX frequency or in the QSX slot.**

A fundamental rule of radio operating is not to call on top of another transmitting station. This also means not to be a "frequency cop" by telling another interfering station that the DX "is operating split, or up 5," or answering another interfering station with "their callsign" or other information. It also means that you don't tune your amp on the live frequency of the operating station, but rather into a "dummy load," - a 50 Ohm resistance load device made for off-air tuning.

- ❖ **I will wait for the DX station to end a contact before I call.**

The best way to ensure that the DX station hears you clearly and responds to your call is to wait until the DX calls for another station. Wait for his/her QRZ or CQ before transmitting to them.

- ❖ **I will always send my full callsign.**

When you respond, make sure you use your full callsign. This facilitates the DX responding with your full callsign and thus ensuring that you are correctly in the log. If you need to correct the callsign you now have that opportunity to do so.

- ❖ **I will call and then listen for a reasonable interval. I will not call continuously.**

We have probably all heard a station transmitting its callsign over and over and not hearing the DX station, who is, in fact calling back to that station's callsign. It almost seems like the calling station is either not hearing the response or is in a trance! Either way, it is holding up all other stations from making their contact. Listen and listen again!

- ❖ **I will not transmit when the DX operator calls another callsign not mine.**

- ❖ **I will not transmit when the DX operator queries a callsign not like mine.**

If the DX comes back to another callsign, be polite, it's not you he wants. Let the other conversation take place. The DX station is not wanting to QSO with you at that moment. Wait until the DX calls for other stations, and then make your call.

- ❖ **I will not transmit when the DX station requests geographic area other than mine.**

A pet peeve of mine is when the DX calls for North America the rest of the world responds. Or when they request calls by "the numbers" (number in one's callsign), and others simply ignore that request and call anyway. This slows down the DX station's ability to work the pileup because they must again state their request. Respect the DX station's request as they are trying to bring some order to the pileup.

and is probably orienting their beam in the direction of the request to focus his/her listening.

> ❖ **When the DX operator calls me, I will not repeat my callsign unless I think they have copied it incorrectly.**

Brevity in QSOs is the mantra of good operation. If the DX station comes back to your callsign correctly, simply give your signal report and a thank you (TU). If it's not correct, then correct it along with your signal report.

> ❖ **I will be thankful if and when I do make a contact.**

> ❖ **I will respect my fellow hams and conduct myself so as to earn their respect.**

The bands belong to all of us. If we want this hobby to prevail and if we want to maintain the privilege of using the bands we must respect each other and share the resource intelligently.

Lessons from DXpeditions – Navassa

Glen Johnson, W0GJ provided some very useful insights into best practices for working DX. His comments appeared in The Grey Line Report, March 2015 immediately after the K1N Navassa DXpedition and were based upon that team's experiences.[iii]

LISTEN to the DX Operator – Are they working simplex or split – and where? Are they calling for any station or a specific continent or callsign number? If you don't know this information you will be calling for naught, and worse – causing DQRM!

Listen in particular for the DX operator's rhythm – do they acknowledge a DXers comeback report and then call QRZ again – or simply call another station?

Identify how the DX operator is working the split – where do they go immediately after working a station? Do they move up or down a KHz or 2? Do they remain on a frequency for one or two more calls? Important information if you are to optimally place your transmit frequency and work the DX in a timely manner.

Know your own radio equipment. Know how to activate VFO B for split operation. Again, if you don't know this you can become the cause of UQRM.

Turn off speech processors and compression and do not overdrive ALC. Distortion makes it difficult to copy on the DXpedition side.

Give your callsign once and then listen. Don't keep calling without listening. If the DX station comes back with your call-sign, it is poor form to repeat your call, as the DX already has it. The DX wishes to make as many contacts as possible, and the additional time taken by repeating callsigns slows the process down. The DX only wants to hear "5NN" or "59" from you. Only repeat your callsign if it needs correction, and then let the DX know it is a correction. Remember, propagation windows and time on the band are limited, and they need to maximize the opportunity for everyone. In DXing, speed and efficiency is the key to success!

DX stations want you to spread out along the split! K1N reported that their highest rates (for any continent) were working the center and far edges of the pileup, where there was less QRM. Weak stations were much easier to work than loud stations in the middle of the pileup. If the DX says, "Listening 200 to 210," 70% of the pileup sits exactly on 200 in an unintelligible din, 25% of the pileup sits on 210 and is almost as bad. 5% of the pileup will be spread out somewhere between 201 and 209, making them very quick to put into the log.

Being loud via high power is not necessarily better! And again, more audio and compression is not the answer. Finding the spot to be heard is the single most important thing you can do to get into the log. DX stations always state that the biggest thrill (I'm sure it's the same on both ends) is finding the lone weak station and getting it into the log.

Listen to the successful DX chasers – operators who make the contact quickly and easily. Take note of their operating practices.

Chapter 3: **Station Requirements**

You really don't need to invest a lot to begin chasing DX. A transceiver capable of working on the standard HF bands (160 – 10M) will do the trick. Couple this with a standard wire dipole antenna or G5RV multiband antenna. If you don't have room for either of those, you might consider a multiband vertical. However, a vertical tends to pick up more noise than other antennas, but they will allow you to work DX if that is all you have room for.

Today, many HF transceivers come with built in antenna tuners allowing a wider operating range on some bands that might have higher SWR at the edges, which is typical of 160, 80 and 40 meter operation. The very popular multiband G5RV antenna, depending upon your bands of interest, may likely require the use of an external antenna tuner. Tuners built into radios can generally only match SWR ratios of 1:3 or less; depending upon the frequency, G5RV antennas can easily have more than that.

Transceivers and Options

Modern High-frequency (HF) transceivers generally cover the 160M through 10M HF bands with many also offering 6M operation. Being able to operate on the 160M through 10M bands is a basic requirement for DXing; having 6M is a nice-to-have capability. Many of the better radios offer numerous options and extras.

The HF bands are often noisy due to atmospheric conditions that create noise (QRN) and from multiple operator transmitted signals (QRM). Therefore, receiver filtering and selectivity are extremely important. I believe the single-most important consideration of any transceiver for DX operating is the receiver's selectivity, the ability of the receiver to tune in a desired signal and reject adjacent and unwanted signals. Also very important is its sensitivity, or ability to receive weak signals. Along with that it is highly desirable to be able to handle a wide dynamic range, which is the ability of handling strong signals both on and off the frequency you are monitoring. The problem is that in some transceivers, if listening to a relatively weak signal with nearby strong signals present, the receiver loses sensitivity and the ability to hear the weak signal. To minimize these problems it is important for a transceiver to accept multiple (roofing) filters and have robust DSP (digital signal processing) filtering and noise reduction capabilities.

Desirable Transceiver Filter Options		
SSB	CW	Digital
1.8K 2.1K 2.8KHz	1.0K 500Hz 400Hz	250HZ 200Hz

Additionally, most modern HF transceivers have dual VFO (Variable Frequency Oscillator) capability, ensuring the ability to operate split frequency – an essential requirement for DXing. This capability allows you to listen on the main or "A" VFO and transmit on a different frequency using the "B" VFO. A separate receive-only antenna connection feature is a nice-to-have feature – especially if you intend to do low-band (i.e., 160M) work.

Transceiver Selection Checklist

- ✓ Frequency range: 1.8MHz – 30MHz plus 50-54MHz for 6-meter work
- ✓ Capable of accepting optional roofing filters as narrow as 200Hz
- ✓ Digital signal processing to 50 Hz built In
- ✓ Receiver sensitivity min. 0.25uV @ 10dB S/N
- ✓ Automatic and manual notch filtering, adjustable noise reduction Better than 60 dB IF Rejection
- ✓ Dual VFOs for working split
- ✓ A built in 2nd receiver is also handy but not essential
- ✓ Transmitter output to 100W adjustable
- ✓ VOX, - Voice Operated Transmit
- ✓ SSB/CW/RTTY/FM and possibly AM Mode operations
- ✓ Transmitter - minimum 40 dB carrier & opposite sideband suppression
- ✓ Transmitter - 3rd order IMD at least 31 dB down
- ✓ Internal CW keyer with message buffers.

Additional Operating Accessories

Accessories that are a must for DXing include a good noise-cancelling headset. Mine serves two purposes. It ensures that I am able to concentrate on the audio from the transceiver without being distracted by extraneous noise. The better ones have a volume control and cushioned sound controlled ear pieces. The second purpose is that the headphone keeps peace between me and my wife! Headsets also come with a mic attached that allows voice activated transmission when the transceiver's VOX circuit (Voice Actuated Transmit) is enabled.

For those who are interested in and capable of sending and receiving Morse code, a key is necessary. Yes, you can chase DX in phone mode only, however, as you gain experience at some point you will appreciate the advantages of using CW. International Morse Code (CW) is still very much used on the Amateur bands and is especially useful when working DX.

CW has the following advantages when chasing DX:

- Permits working DX pileups faster.
- Facilitates more accurate exchanges of call signs and signal reports.
- Reduces international language difference difficulties.
- Occupies less band space than a SSB signal, thus reducing conflicts between DX stations and the general Amateur community (it is possible to hold 10 different QSOs in the band space required for one SSB QSO).
- CW gets through the band noise very efficiently, and is especially useful in low power (QRP) operations.
- CW requires some operator prowess and is a specialty skill one should be proud of perfecting.
- CW permits people with disabilities that prevent voice communication to operate and make contacts on the radio.
- A CW signal can have a 10dB advantage over a SSB signal. I have had success in completing QSOs with 100W of power in CW that would have required 1KW in SSB mode.

Some DX stations only work CW. If a DX station using CW is a rare entity, not being able to work CW will be a problem. Making things a bit easier, CW can be sent from a computer keyboard through a software interface. My interface is built into my logging program. This is a common logging program feature, available in almost all commercial logging programs and some freeware ones as well. To be effective in DXing with CW, however, you need to be able to copy some CW – calls, signal reports and the common Prosigns.

A third mode of operation, digital, requires an interface of some sort. Today, many higher end transceivers have this built in. DX digital modes include radio teletype or RTTY, PSK 63 & 125, and BPSK 63, and this is constantly evolving as new technology is constantly being developed. I prefer RTTY and use an external digital interface to my transceiver. Many hams prefer one or more of the multitude of software supported digital modes.

A 24 hour clock is also a necessity. DXing requires that the station operators on both sides of the contact be in sync as far as the time of the contact is concerned. Therefore we use a 24 hour time standard adjusted to Coordinated Universal Time (UTC). DXers will know how many hours to add or subtract from their local time to report the time in UTC. Your logging program will automatically record a QSO in UTC, once you set the program up with your local grid square or longitude and latitude.

Finally, you need some way to log contacts. Certainly you can keep a paper log – but this will be very cumbersome, especially if you need to look up past contacts quickly. Most of us use computer logging. Numerous logging products are available, both freeware and purchased, offering varying degrees of bells & whistles. This is discussed in some detail in Chapter 5.

Antennas and Towers

A fundamental reality of DXing is that the antenna makes all of the difference. A sophisticated transceiver with all of the bells and whistles and a 1.5KW linear amplifier is of little value if you can't hear the DX station. Put your disposable income into your antenna, if at all possible. Dipoles and G5RV wire designs were previously mentioned. These are good choices, and only require a moderate amount of space in your back yard. Space permitting, you might want to consider larger wire designs (loops, inverted L, Sterba curtains) or Yagi's and other beams. Wires can be strung up in trees, and/or supported by hanging a portion from your house. Many hams shoot up wires with slingshots or bows and arrows. Most local clubs have someone who has a talent for shooting up ropes and supports for wire antennas. Also, there are several good books on wire antenna designs for DXing. I suggest obtaining and reading Press Jones' N8UG, *"Wire Book V"* [iv] available at **www.thewireman.com** .

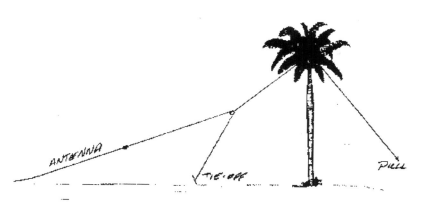

Raising wire antenna using a pulley and tie-off
From: http://www.radioworks.com/ninstallant.html

Directional beams constructed of extruded aluminum tubing that can be raised up to 50 ft. to 65 ft. or higher are the optimum operating antennas for chasing DX. Beams require a structure for support – usually a tower. Towers range from 35 FT to 70 FT or more. When considering a tower, in addition to ground mounted towers consider a roof mounted design if you are unable to ground mount one.

Directional beam antennas of the multi-element Yagi design can be cut for a single frequency band, or multiple bands. There are also log periodic antennas that cover a range of bands. Many hams prefer to use multiband antennas for the convenience they offer. Directional antennas require a rotator and controller. An excellent reference for HF antennas is *"HF Antennas for Everyone"* by Giles Read, G1MFG[v], which provides a collection of information on a wide variety of antennas: horizontal, vertical, loop and stealth designs.

Connecting the antenna to the transceiver requires some form of feed-line. Ideally it's best to use coax cable of good grade and minimum signal loss (perhaps LMR 400). Runs should be as short as possible, and made with very good coax to minimize loss. For wire antennas such as a loop or G5RV design, consider open ladder line terminated in a balun. Again, read up on these configurations in a good antenna reference book.

One can't discuss antennas and feedlines without touching on station grounding to complete the circuit. DXing requires good station grounding and I suggest you read up on this as well. I have my tower grounded with three 8 FT ground rods – one at each leg. All of my equipment in the shack is grounded and terminated at an 8-FT ground rod outside of the shack.

Finally, an antenna tuner will be needed if you are using a wire antenna designed for multi-band use, or a beam that is not resonant across all of the bands you want to work on. I suggest investing in a high-power tuner so that you can go QRO (high power) at some point.

Lightning Protection

When installing antenna systems, you should always think about mitigating the possibility of lightning strikes. Not only can lightning be personally dangerous, it cause damage to your radio equipment or home. Perhaps nothing can completely prevent damage from a direct hit, but quite a bit can be done to mitigate the effects of nearby strikes.

To give some measure of protection to your radio gear, the simplest thing you can do is unplug the power and antennas from your transceiver and antenna tuner. However, this does not protect your home and equipment from lightning induced surges that might come down your antenna feed-line and arc over to something in the shack.

The best protection comes from shunting any atmospheric induced energy to earth with a good ground system. The emphasis is on good the connection needs to have a large ground strap leading to an effective ground rod. Typically gas discharge devices, sometimes called PolyPhasers are inserted in series with the antenna feed line to shunt surges to ground. If a sufficiently high voltage appears on the antenna feedline, the device discharges, providing a relatively low impedance path to ground. For this to work, the protection devices are directly fastened to a large single point ground plane, which is then connected to earth ground with a wide ground strap, usually of an inch or more in width.

If you are a member of the ARRL, you can download a three part article titled *"Lightning Protection for the Amateur Station"*, which appeared in the June, July and August 2002 issues of QST magazine. Another good article is the *"Ham Radio Station Protection, Technical Note"* available from Smiths Power. These comprehensive articles will take you through the entire process of assessing your needs, the types of protection devices available and how to implement an effective station earth ground system. Both articles will readily come up if you do an Internet search for "Ham Radio Station Protection."

QRO / High Power (or not)

DXers love linear amplifiers! It's a guy thing, I guess. More power – more muscle. To a degree this is true. In a pileup for a rare DX station in less than optimum band conditions, the RF signal heard on the DX station's end with the highest S–unit rating will generally be heard first. Therefore, a linear amplifier will help. That does not negate lower power signals strategically located in the spectrum split being heard.

Over time you will gain experience leading you to make a decision on an amplifier. While not immediately necessary, to the extent practical a linear amplifier will be a nice addition to your shack. It does not have to be a "full gallon and a half," (1500 Watt); 500W amplifiers now proliferate the market and at relatively reasonable prices. Since the difference between a 500W and 1500W amplifier on the other end of a pileup is less that an S-unit in signal strength, 500W amplifiers can be a good value. When you decide to acquire one, there is plenty of good information out there to help you determine what kind of amplifier makes sense for you. eHam.net's product reviews on liner amplifiers is an excellent place to begin your study. **www.eham.net/reviews/**

Amateur Radio World Map & Reference Materials

DXing requires good reference material to be close at hand. If you are going to chase worldwide radio stations, you need to know where these countries or entities are physically located. A good ham radio map, with ITU zones delineated, is essential. Hang it where you can easily reference it.

Additionally, the ARRL DXCC List is handy, as you will find that numerous international radio prefixes exist, and some are not very common. While most logging programs quickly provide alternatives for an assigned prefix, sometimes you have to help out.

A very useful reference document that I use almost on a continuous basis is my personal beam heading listing. This I obtained from the *North Jersey DX Association* by inputting my QTH's grid square. You can create and print your own at: **http://njdxa.org/dx-tools/beam-headings.php**[vi] As an example, a small portion of my customized Beam Headings Chart is shown below. The full chart shows all prefixes.

	The North Jersey DX Association Presents Your Personalized DXCC Countries List & Beam Headings							
	Beam headings prepared for, Jeff, K1ZN Headings centered on Senoia, GA - Your Latitude 33.28 Your Longitude 84.59							
Prefix	Country	Short Path	Long Path	Miles	Latitude	Longitude	Continent	CQ Zone
1A	SMOM	51	231	5,047	41.9	-12.4	EU	15
1S	Spratly Is.	337	157	9,327	8.8	-111.9	AS	26
3A	Monaco	50	230	4,760	44.0	-7.5	EU	14
3B7	Agalega/St. Brandon	66	246	9,459	-10.0	-56.0	AF	39
3B8	Mauritius	79	259	9,942	-20.3	-57.5	AF	39
3B9	Rodriguez Is.	73	253	10,251	-19.7	-63.4	AF	39
3C	Equatorial Guinea	86	266	6,410	1.8	-10.0	AF	36
3C0	Annobon	89	269	6,164	1.3	-5.4	AF	36
3D2	Fiji	259	79	7,269	-17.0	-178.0	OC	32
3D2/C	Conway Reef	256	76	7,610	-22.0	-175.0	OC	32
3D2/R	Rotuma Island	264	84	7,147	-12.0	-177.0	OC	32
3DA	Swaziland	102	282	8,645	-27.0	-31.5	AF	38
3V	Tunisia	56	236	5,121	36.3	-10.2	AF	33
3W/XV	Vietnam	344	164	9,295	10.8	-106.7	AS	26

Know the CQ & ITU Zones

The entire surface of the world is partitioned for the purpose of radio communication into subdivisions called zones. Two zone systems are in general use by Amateur Radio operators: CQ and ITU. ITU zones were created to aid worldwide communications and there are 90 of them. DX stations will indicate the ITU zone of the operation when they announce their intention to work as a portable from a lesser known location and entity, thus permitting the rest of us to identify the location in the world and set our beam headings. The alternative CQ system of dividing the world into zones was developed as a basis for the CQ Magazine sponsored awards program. Both systems are in common use.

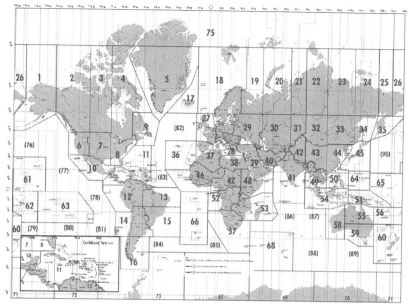

World ITU Zone Map
Both maps courtesy of Icom America

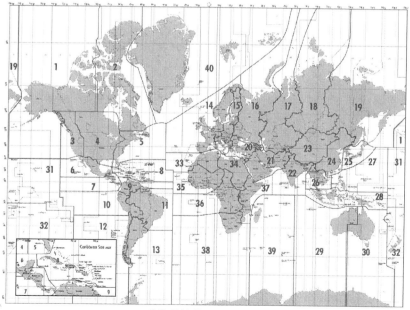

CQ DX Zone Map

Using Contests to Catch DX

Contesting helps build operating skill and contributes to increasing DX station counts. Contests on the Amateur bands are a form of radio sport. Contests involve hams going on the air with the objective of contacting as many other stations as possible in a fixed period of time. Contests vary by most every Amateur Radio variable imaginable – mode of operation (CW-RTTY-Phone); Band (i.e., 60M - 10M, VHF – 6M or 2M); Country/Entity (i.e., 4X; LA); US Counties; Islands–on–the–Air, etc. Basically, these are organized activities where stations seek to make two-way radio contacts with each other over a specific period of time, using the specified: modes of operation, signal reports and station identification.

Contests proliferate – there are contests on the air every weekend. They provide a recreational opportunity for playing radio while honing operating skills and testing your equipment setup under contesting conditions. Contests are an excellent opportunity to work DX stations that come on the bands at the time of the contest for much the same purposes.

For example, there is a *"Worked all US Colonies"* contest sponsored by the U.S. Original 13 Colonies Group **http://www.13colonies.info/** Stations in the U.S. and elsewhere will go on the air for a specified period of time (i.e. for this contest from Tuesday 1300Z until Sunday 0400Z), to work at least one of several specially identified stations in each of the thirteen original U.S. colonies. Having achieved a "clean sweep" of all 13, the contesting station can then apply for an award certificate.

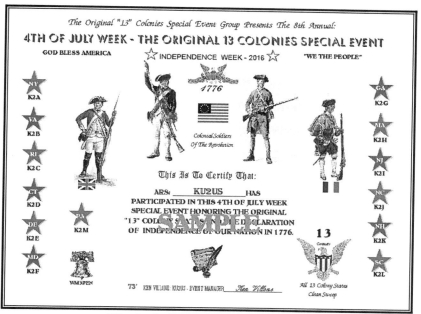

Worked all US Colonies Award
Courtesy of www.13colonies.info

While I am not an avid contester, I pay attention to the contest calendars and use them as an opportunity to seek out DX, not necessarily ATNOs (*all-time-new-ones*), but for band fills – new entities on bands other than the ones that I already have worked and confirmed an entity on. For example, I am presently working towards DXCC on 160M and 6M. When there is a contest on one of these bands, the band comes alive, presenting a golden opportunity for picking up new entities on that band. So, contesting is an activity that a DXer should pay attention to.

As I am working towards DXCC on 160M, I am using the contests on 160M to pick up as many countries and entities as I can; many DX stations come into the contest specifically to work new DX. Don't overlook these opportunities for working new DX.

To participate in a contest and hone your DX operating skills, and perhaps log an ATNO (*all-time-new-one*), all you need to do is identify a contest that interests you and research the exchange requirement. In the case of the colonies contest mentioned above, the exchange is your call sign, name, the signal report (RST) and your state, province or country. I have included a sampling of contests that run annually, and by far it is only a sample. Also, QST Magazine publishes a monthly calendar of contests in their *"Contest Corral"* section.

Some contests include:
- ARRL 10 Meter Contest
- IARU HF World Championship
- International DX – CW
- CQ Worldwide 160M
- RTTY Roundup

Chatham Island, IOTA QSL Card

Sample List of Contests and Objectives		
Contest Name	Description	Scope
Helvetia Contest	Worked all Swiss cantons	1-day, all bands & modes
Straight Key Weekend Sprintathon	CW straight key operator competition for maximum number of unique calls completed – once a month event	1-1/2 days, all bands + 6M, CW straight key
Ten-Ten Spring CW Contest	10 Meter CW competition to log as many member 10-10 numbers / stations as possible on CW	1-day, 10M, call sign, 10-10 number & state
CQ WW WPX Contest	An international contest to work as many call sign prefixes as possible. A good contest for DXers	2-day, all HF bands, RST plus serial number (first contact – 001, second contact – 002, etc.)
Indiana QSO Party	Annual contest sponsored by an Indiana radio club(s) to facilitate Indiana stations to work as many stations outside of the state as possible – good contest for county hunters	1-1/2 day, RST, state or province, DX – RST only

You might also check out the contest calendar maintained by WA7BNM on the web at: **http://hornucopia.com/contestcal/** Contesting.com is also a good resource for daily information about upcoming contests.

DX Nets and Call Lists

A net is a gathering of hams on the band for a specific purpose. In the case of a DX net, the purpose is for aspiring DXers to come together to work new countries. A net is a controlled gathering, wherein there is a net control station and often some assisting stations that serves as a moderators. All traffic passes through the net control or moderator(s). DX stations are solicited by the net control operator as they gather and sign-in on that frequency. Next, the net control operator then calls for stations wanting to work those DX stations to call-in and get on his list.

One-by-one the net control then calls upon those on the list to call one or more of the DX stations. The net control station then verifies a good contact or QSO on both sides.

A positive aspect of a net is that it often mitigates language limitations – DX operators who have limited English capabilities and DX chasers who only speak English. The net control operator understands these limitations and helps moderate.

Naturally, using a pre-solicited list avoids a pileup situation for new DXers. You wait your turn and get an uninterrupted try at the DX you are seeking. Only your equipment setup and propagation stands between you and a good contact.

I found that at the beginning of my quest for DXCC, the net process helped me avoid freelancing on the bands and hunting for new DX stations. It gave me an opportunity to build my list of acquired countries and motivated me to get into the "open hunt competition" of pileups and band chasing. An added benefit is that in some cases, net control stations also serve as a QSL clearinghouse, making the task of getting a confirmation card that much easier. Years ago, on the 247 DX net (20M), Alan, WA4JTK (SK)[vii] brought together a great number of good DX stations, was a good net control operator, and performed the QSL manager task well. It was on this net that I was able to achieve my first 100 entities and DXCC.

Recently I was listening to the HHH net on 7.190MHz around 0730Z. The net had check-in DX stations from VK, KH6, VE and CO, all booming into the east coast of the USA. As a very efficiently run net, it was easy to work DX stations on the 40M band.

Here are a few pointers when checking into a net. First listen to the net and learn how it operates. Usually the net control(s) will give a preamble discussing net practices: what to give on check in (i.e., name, QTH, how many calls you can make per turn, what to give on contact (signal report and name), and any other rules such as not relaying information unless asked to do so. If you use a beam antenna, turn it to the net control for check in, and when the time comes towards the DX you wish to work. Above all, if a QRM'er appears on the frequency – ignore that station and take direction from net control.

Best of all, DX nets are on the bands at a fixed time and fixed location, making operating time planning easy and possible. Some long standing DX nets are:

Long Standing DX Nets			
Net Name	Frequency	Start Time	Information
Triple-H Net	7.190 MHz	0700Z	www.hnhnet.net
247 Net	14.247 MHz	2100Z	
W7PHO Net	14.245 MHz	1400Z	
ANSA Net	14.183 MHz	0515Z	
Southern Cross DX Net	14.239 MHz	1100Z	
10-10 International	28.380 MHz	1800Z	http://www.ten-ten.org/
OMISS Net	Multiple	Depending on Band	http://www.omiss.net/facelift/index.php
1865-tomato.net	1.865 MHz	0430 – 0600Z	http://www.1865-tomato.net/Home_Page.php

Remote Operating

You are probably familiar with portable ham radio operating. You have a hand-held radio and perhaps a portable HF rig you have taken out camping, hiking and to picnics. You probably are a mobile ham radio operator with a mobile VHF radio in your vehicle. Remote operating for DX is not all that different – except that in remote DX operating, you would have a full station, inclusive of a tower and antenna in a location other than where you are physically present.

For instance, a ham working DX from San Diego with their transmitter and antenna located in Boston has become common place. Today, we are seeing more and more transceivers that can be remotely controlled via the Internet. This situation is perfectly legal under Part 97 of the FCC rules and regulations, as long as the operator is legally licensed for the band they are using and in control of their transmissions.

At the time of writing this book, Amateur Radio commercial ventures have appeared that provide time rentals of remote stations (full transceivers, linear amplifiers, towers and antennae, and interface software for control purposes).   Thus, Amateurs now have the option of building their own remote location towers and interfacing their rigs via the Internet – or renting fully operational stations elsewhere in more favorable geographic locations for the purpose of operating DX.

Chapter 4: **Propagation**

When we speak about propagation we are referring to radio wave propagation or the behavior of radio waves when we transmit them from our shack with the intention of directing them through the atmosphere to a distant location. The study of the factors influencing propagation is a rather complicated science. Fortunately a lot of work has been done in this area and by referring to published propagation predictions we can make decisions about the best choice of frequency bands for desired destinations.

Propagation Basics

Some knowledge of the ionosphere and how it is affected by solar radiation, time of day and year will aid in understanding radio wave propagation and how it affects our ability to hear DX, and the ability to transmit back to the DX station. The ionosphere plays an important role in DXing because it influences radio wave propagation and the flow of radio waves across the earth.

The ionosphere is a region of the atmosphere approximately 37 miles to 370 miles above us. Basically the ionosphere can be thought of as a shell of electrons and electrically charged atoms and molecules surrounding the earth. The ionosphere is primarily energized by ultraviolet radiation from the sun. This radiation disrupts the gasses in the upper atmosphere, freeing electrons – hence causing ionization.

Solar radiation has a major influence on skip (the propagation of radio waves reflected back to earth from the ionosphere). The amount of solar radiation is highly correlated to sunspots. The more sunspot activity – the more solar radiation that is generated – the more ionization and the more long-distance communication will be enhanced on our bands.

You have probably heard hams talk about regions of the ionosphere. Well, here's the scoop on that. For our purposes there are several regions of the ionosphere that affect radio wave propagation. These are:

> ➢ D - Region – the region closest to ground about 30 to 60 miles up. It is of little value to DXers during the day, as it is least

ionized and completely absorbs sky wave signals in the M (80-160 Meters). At night the D-Layer ionization disappears. The characteristics of this layer of the ionosphere makes short hop skip possible. As an avid 160M DXer I have learned that as a result of the D-layer, signal travel distance is affected by both topography and power of the signal. Sandy soil is no good for signal travel, whereas moist soil is very good, and the sea is an excellent medium. Also antenna choice is important, vertical antennas are best for D-region ground waves. Pre dawn DXing and DXing after dusk is a must on the low bands. Long distance paths must remain in the darkness for long distance communication to happen.

- E – Region – Medium Frequency and High Frequency wave absorption is common in this region. E-region supports single hop skip to about 1200 miles.

- F-Region – The F-region is the ionosphere region most significant to HF DXing. Located approximately 60 to 70 miles up. During daytime hours we speak of the F-region as being in two sub layers – F1 and F2. The F2 sub layer is said to be responsible for long hop skip to 2500 miles. This sub layer supports daytime DX.

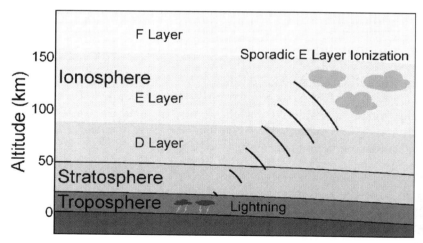

Regions of the Ionosphere

Radio propagation describes the behavior of radio waves as they are transmitted from our antennas to intended stations somewhere around the world. Radio waves are a form of electromagnetic radiation and are subject to the effects of a number of forces or phenomena such as absorption, reflection, refraction, etc.

The electromagnetic spectrum that we are concerned with is:

VLF	Very Low Frequency	3 – 30 KHz
LF	Low Frequency	30 – 300 KHz
MF	Medium Frequency	300 -3000 KHz
HF	High Frequency	3 -30 MHz
VHF	Very High Frequency	30 – 300 MHz
UHF	Ultra High Frequency	300 MHz -3 GHz

Radio signals can travel in several forms: direct line-of-sight, ground wave, or as a sky-wave. We are familiar with line-of-sight wave travel as we have all used VHF hand-held transceivers in FM mode. With it we are able to communicate effectively to the limit of the power of the transceiver - from our hand-held directly to the other receiver without any form of skip being involved.

We have probably experienced ground wave forms of propagation as well, if we have been in QSO's with a close-by ham on the HF bands, or perhaps for an early morning net, or as part of an ARES group, locally. This form of propagation is very dependent on terrain and composition of the ground, locally.

Sky wave propagation is the form of propagation dependent on the ionosphere and the times of day/evening at both ends of the communication. "Skip" as it was called when I first got into the hobby, is the most erratic, and the one we dwell on most as DXers.

Propagation Varies by Band

The following table summarizes general characteristics about the bands used to work DX. As shown, these characteristics vary depending upon the time of year, time of day and the amount of sunspot activity.

Propagation Characteristics by Band

Band	Characteristics / Uses
160M 1.8 – 2.0 MHz	Called the 'top band,' this band is best used after sunset and just before sunrise. After sunset in the winter months, it's possible to work stations thousands of miles away, but summer months DX typically will only be within hundreds of miles. The band is noisy with static crashes in the summer months. Hope for rare sunspot activity for this band to be at its best. Increased transmit power (QRO) is a must on this band.
80M 3.5 – 4.0 MHz	This band's characteristics are similar to 160M, with DX capability for greater distances at night and in the winter months. This band is less susceptible to sunspot cycles and thus more predictable and reliable. Like 160M, DXing on 80M is challenging. A good late night DX band.
40M 7.0 – 7.3 MHz	This is a "workhorse" band for the Amateur Radio community. It is crowded with hams as well as international shortwave broadcast stations. Winter day DXing you can expect a maximum range of about 500 miles, but in the evenings you can expect intercontinental ranges. Summers less so, with day ranges of up to 400 miles and eves of 1,000 miles. Sunspot cycle effects are moderate on this band. Nets use the band extensively. I find that this band always seems to be able to support long distance communication to somewhere in the world.
30M 10.100 – 10.150 MHz	This band is restricted in the U.S. to CW & RTTY. Characteristically, it is much like the 40M band, but offers a bit more DX range, maybe 1,000 miles or more during the day, and even more in the evenings.
20M 14.000 – 14.350 MHz	This band has been referred to as the "Times Square" of the ham bands, as everyone migrates to it. Oftentimes frequency space is at a premium. At the peak of the sunspot cycle the band supports international / intercontinental DXing around the clock. At the bottom of the cycle, openings will be short and often rare. However, local (50-75 mile) communication will only be ground wave. I find the band is generally open for DX at some point during the day. Check into a 20M net to make an assessment of the band.
17M 18.068 – 18.168 MHz	Characteristics similar to 20M. Being less crowded makes it somewhat easier to make contacts faster. This is one of my favorite DX chasing bands.

15M 21.000 – 21.450 MHz	The 15 Meter band is somewhat like the 20M band in that it is popular with the DX community. This band's propagation is perhaps more predictable that 20M – more influenced by sunspot activities. It is quieter than 20M in terms of atmospheric noise. At peak sunspot activity more signal distance can be anticipated. It is a good band for summer DX work during periods of low sunspot activity.
12M 24.890 – 24.99 MHz	12M is very dependent on the sunspot cycle to support radio wave propagation. However, at the height of the cycle, this band can make for exciting DX around the world. In just a few years I worked DXCC on 12M barefoot!
10M 28.000 – 29.700 MHz	Ten meters straddles HF & VHF, being the upper edge of HF and bottom edge of VHF. This was my first DX band, while still studying to advance my license class. While the band is the most affected HF band by sunspot activity, when it's open, it is open for hours. Ten meters exhibits characteristics of both HF & VHF. I found that it is best to use beacons for assessing 10M propagation, as it does shift in direction throughout the day. Sporadic-E openings also occur in summertime on this band making for some exciting DX.
6M 50.000 – 54.000 MHz	The magic VHF band as it is often referred to, 6M DX is in a class by itself. Patience is what is needed and close beacon monitoring for openings that usually occur as summer approaches. With three decades of DX behind me, I have 54 entities confirmed on 6M. Patience!

The best way to test out the bands is to get on and listen. Don't forget the reverse beacon network (to be discussed at the end of this chapter) as a way verifying signal propagation to distant locations.

What do Propagation Forecast Numbers Mean?

You most likely have heard DXers talk about sunspot activity numbers and indices of one kind or another. You have likely seen these numbers listed on various DX spotting web sites. What do they mean, and how can we use them to our advantage? At the time of this writing, we are at the bottom of Solar Cycle 24 – perhaps the weakest that has ever been recorded. Solar cycles tend to run in 11 year cycles, so they take a while to play out.

```
Solar-Terrestrial Data
05 Sep 2015 1516 GMT
SFI:91      SN:   36
A-Index: 20
K-Index: 2
X-Ray: B1.1
304A: 128.5 @ SEM
Calculated Conditions
  Band      Day   Night
80m-40m:   Fair   Good
30m-20m:   Good   Good
17m-15m:   Fair   Fair
12m-10m:   Poor   Poor
Signal Noise:    S1-S2
Click to Install Solar
Data On your Web Site
http://www.n0nbh.com
Copyright Paul L Herrman 2010
```

Courtesy of Paul L. Herrman, N0NBH

Let's decipher the above solar data chart that we found on DX Summit http://www.dxsummit.fi/#/ Using the following table we can relate the above readings to the effect they have on radio propagation.

Indicator	Reading on Chart	Meaning
Solar Flux Index (SFI)	91 60 - 70 = No good 80 – Good 90 – Better 100+ Best	On a scale of 60 (no emissions) to 300, this measures the total radio emissions from the sun.
A – Index	20 1 – 6 = Best 7 – 9 = OK 11 or More = Bad	Indicates the overall geomagnetic condition of the atmosphere.
K – Index	2 5 = geomagnetic storm 1 = calm	Quantifies disturbances in the horizontal component of the earth's magnetic field.

SN	36 The high reading in the current Cycle 24 was 296 in April of 2014	Current sunspot numbers as determined by NOAA Space Environment Center.
X-Ray	B – 1.1	X-Ray is the background flux from the sun. M & X solar flares impact D – Region leading to radio blackouts. A – B = Normal C = Moderate M, X = Strong
304A	128.5	The 304A number is an indicator of the radiation from the sun, and is a good indicator of F-2 Region propagation and thus higher band activity.
Signal Noise	S-1 to S-2	This is an estimation of signal noise levels heard on the bands due to atmospheric conditions.

The "Calculated Conditions" section of the solar data chart provides estimates of day and evening propagation conditions by band which can be used to your advantage in deciding which bands to work or destinations to seek.

Working Greyline

The term Greyline refers to the area occurring along the sunrise and the sunset twilight zones. Radio waves that travel along these two zones tend to have enhanced propagation due to the absence of D-layer absorption. The F-2 layer supports these signals. Greyline on 160M in the early morning can last 15 minutes or less; 40M greyline can last up to an hour. Many ham radio programs can display the current greyline superimposed on a world map as it moves across the globe. On the web, DX.QSL.NET provides a continuously updating greyline map at
http://dx.qsl.net/propagation/greyline.html

Long Path

Sometimes propagation conditions are such that a station's signal is stronger when you point your beam "long path" or in the direction opposite the direction indicated on your beam heading chart. As I was developing this chapter I was attempting to work an Indonesian island

and had my beam pointed at the station, (YB is 336° direct or short path for me). I did not hear a signal. Then I noticed that other nearby stations were spotting him LP. I turned the beam to 156° and there he was!

Skew Path

Skew path is any path by which the signal is offset by less than 90 degrees from its true Great Circle bearing. Skew path is very intermittent and seems primarily related to periods of ionospheric disturbance. An article by David Craig describing several different skew paths from the East Coast can be found at http://www.qsl.net/n3db/propagation/Dec31a.htm and is informative as to the nature of skew path propagation.

Sporadic-E Propagation

Sporadic-E propagation is propagation of radio waves in the Lower E region of the ionosphere wherein the radio signal is reflected by small patches of unusually ionized atmospheric gasses in that region. Six-meter radio activity is very much affected by Sporadic-E. TEP, or transequatorial propagation, a form of VHF band intercontinental communication during high sunspot activity, usually occurring late afternoon and early evening, provides great opportunities for working DX on 6 meters.

Propagation Prediction Resources

There are a number of good propagation prediction programs and resources available to us. One such resource is provided by the American Radio Relay League. Their web page hosts a propagation resource page at http://www.arrl.org/propagation/ The page then directs you to a couple of solar data web sites and also provides HF Prediction charts based upon your general location in the United States.

Beacons

Beacons are radio stations that operate on a fixed frequency and continuously transmit a signal – usually CW, for the purpose of others to listen and try to hear, thus providing an indicator of propagation on the band. If you know the beacon's location, then you will know how the propagation is behaving on that frequency to that location.

The *Northern California DX Foundation* (NCDXF) and the *International Amateur Radio* Union (IARU) constructed and operate a worldwide network of HF beacons. Using Morse code, the beacons transmit their callsign followed by four 1-second dashes at 22 wpm. The callsign and first dash are transmitted at 100 watts, followed by dashes of 10 watts, 1 watt and 100 mw. All of the beacons transmit every 3 minutes on a staggered schedule, on 20, 17, 15, 12 and 10 meters. Even if you can't copy 22 wpm, by knowing the exact current time and the beacon transmission schedule you can tell which beacon is transmitting at any given instant.

NCDX/IARU Beacon Transmission Schedule
Times shown starting on the hour. Beacons repeat every 3 minutes.

Call	Location	14.100	18.110	21.150	24.930	28.200
4U1UN	United Nations	00:00	00:10	00:20	00:30	00:40
VE8AT	Canada	00:10	00:20	00:30	00:40	00:50
W6WX	United States	00:20	00:30	00:40	00:50	01:00
KH6WO	Hawaii	00:30	00:40	00:50	01:00	01:10
ZL6B	New Zealand	00:40	00:50	01:00	01:10	01:20
VK6RBP	Australia	00:50	01:00	01:10	01:20	01:30
JA2IGY	Japan	01:00	01:10	01:20	01:30	01:40
RR9O	Russia	01:10	01:20	01:30	01:40	01:50
VR2B	Hong Kong	01:20	01:30	01:40	01:50	02:00
4S7B	Sri Lanka	01:30	01:40	01:50	02:00	02:10
ZS6DN	South Africa	01:40	01:50	02:00	02:10	02:20
5Z4B	Kenya	01:50	02:00	02:10	02:20	02:30
4X6TU	Israel	02:00	02:10	02:20	02:30	02:40
OH2B	Finland	02:10	02:20	02:30	02:40	02:50
CS3B	Madeira	02:20	02:30	02:40	02:50	00:00
LU4AA	Argentina	02:30	02:40	02:50	00:00	00:10
OA4B	Peru	02:40	02:50	00:00	00:10	00:20
YV5B	Venezuela	02:50	00:00	00:10	00:20	00:30

Reverse Beacon Network

Alternatively, a reverse beacon network is a system whereby you transmit a signal out into space on a given frequency and those stations hearing you respond by sending out signal reports. *The Reverse Beacon Network Project* has developed just that, and it is a wonderful tool for assessing band conditions at a particular day and time. These are not predictions; it is what is actually occurring at the moment.

This project (RBN) relies on volunteer stations to set up their station on selected frequencies to participate. The Reverse Beacon Network decodes and collects multiple CW and RTTY signals from over a hundred receiving stations all over the world. These reports are sent to a central server from where the spots can be distributed and posted on web pages, which is where you can see your results when you transmit. A sample of what is currently being received on the HF bands can be seen at **http://www.reversebeacon.net/dxsd1/dxsd1.php?f=20**

A good introductory slide presentation explaining this technology is available at:
http://www.kkn.net/~n6tv/N6TV_Dayton_2013_Using_the_RBN.pdf

VOACAP, HF Propagation Prediction Software

As you get more immersed in the art and science of DXing, you will appreciate tools that make the venture more successful. One such tool is VOACAP – *The Voice of America Coverage Analysis Program*, originally developed by Uncle Sam for the Voice of America. By inputting your QTH, antenna configuration and power output, the program can be used to predict the quality of point-to-point radio transmissions over a range of frequencies. It is still in refinement and is a public domain program under the aegis of the *National Technical Information Service* (NTIS/ITS). You can download the software and user's manual at: **www.VOACAP.com** Alternatively, you can easily run a simplified version of the program on-line at the same web page.

Chapter 5: **Spotting Clusters and DX Publications**

Spotting DX is an important part of hunting DX for me. In the mid-1980's, when I got serious about DXing, I joined a group of local hams and DXers who were alerting each other of DX stations operating on the bands using 2M packet radio as the communications medium. These electronic bulletin boards were called DX clusters, and became an invaluable asset to hams for finding DX on the bands, noting their callsign and frequency of operation. Amateur Radio operators that are logged into a cluster can see stations spotted by others in real-time, or post their own observations. Over time, local packet clusters became linked to other packet clusters both regionally and internationally, dramatically increasing the number of spots being posted. We are now in the digital age and packet clusters have migrated to the Internet. If you have a computer and Internet access, observing DX spots that have been posted, or leaving posts of your own is about as easy as it gets.

Online Spotting Programs & Operating Resources

DX Summit.
DX Summit is an online international DX station spotting program with search history capabilities that is operated by Radio Arcola, OH3X. This simple to operate and robust spotting program has the ability of setting numerous search filters: DX entity, band, mode, IOTA number and callsign. For example, when I was trying to spot Minami Torishima, JD1/M, I was able to set a filter to search for his callsign so that whenever his call was posted by someone, worldwide, I was instantly alerted. DX Summit also stores historical data, allowing you to search for a station that had been spotted in the past, by date and/or time. To keep abreast of the latest happenings, DX Summit also links to Bernie's, W3UR, "*The Daily DX*" for informational updates and brief notes.

The following image was captured from the DX Summit web page. Notice the simplicity and effectiveness of the spots. The filters that were used to limit the number of spots are shown at the top of the page. Below that you see: the spotter reporting the DX, the date and time of the report, callsign of the station being spotted, and the frequency of operation. http://www.dxsummit.fi/#/

Sample DX Summit Spotting Page
Courtesy of DX Summit

Note: Filters were set to limit the display to phone mode DX spots on the 3.5, 7 and 14MHz bands.

QRZ.com.
QRZ.com is a multi-database site that includes a callsign search engine, DX spotting cluster, plus multiple related databases and resources. The site includes a DX country atlas, a QSL routing database, swap and trade listings, specialty forums and news of upcoming hamfests and conventions, general interest, callsign history and availability databases, license preparation resources, grid mapper, and more. As an Amateur Radio resource, this is about as comprehensive as it gets. Users will need to establish a user name and password to use the site,

which is free. Subscribing and paying a fee, however, entitles you to full use of all of the site's additional capabilities. See: **www.qrz.com**

eHam.net.
Similar in nature to QRZ.com, this site also has a spotting cluster with nice filtering capabilities. I have found that the QSL manager and routing database is kept up-to-date. The site has a very useful product review service, and its propagation resources and contesting support and links are also handy. If you are looking for a good place to buy and sell ham radio gear, eHam's classifieds section is well subscribed to. Like QRZ.com, users must register to fully use the site; registration is free. See: **www.eham.net**

DX Clusters - Use & Etiquette

All of these online clusters have the ability to accept spots, which are postings by any ham monitoring the bands. However, there are rules of the game that are important to understand.

- Typically, the station that works the DX will type in the necessary information for others to benefit: station call, frequency the station is operating on, split if any, and possibly mode of operation and other pertinent information.

- It is frowned upon to post spots unless you work them.

- It is also frowned upon to use the clusters as a medium for extraneous comments.

- Do not self-spot to announce that you are on a particular band or frequency.

- You should not "thank" a DX station for picking you up.

- You should not comment on another station's operating practice.

ClubLog.
ClubLog was built by Michael, G7VJR. It is an online tool that supports DXing activity in a wide variety of ways, including: running reports showing DX entities you have worked, details about when you worked them, which ones you need, and where you rank in comparison to other ClubLog users relative to the number of DXCC entities and CQ Zones contacted. In addition to hams establishing their own individual account, DX clubs can establish an account for member

tracking purposes and for enabling member contacts to be contributed to the club's overall totals. Once I established my individual account, I was able to select my club (SEDXC) affiliation in the ClubLog program so that my individual statistics contributed to my club's statistics.

To make use of ClubLog you must register, which you can do for free. When registering, however, be mindful that the services provided by the ClubLog web site are supported by voluntary donations. You can register at **https://secure.clublog.org/loginform.php**

ClubLog provides an active DXer with the following tools:

- Personal DXCC reports. Much like the ARRL Logbook of the World (LoTW)™ program, I am able to see DX entities that I have worked and confirmed (based upon my log upload information), verifying and analyzing bands and modes per entity.

- Perform a detailed analysis of your log by: needed DX, confirmed DX, and DX by band and mode. You can do the same analysis for your club affiliation as well.

- Perform these same analyses by year of operation.

- Search your log for any kind of information pertaining to the DX. For instance, I can do a fast online search to see if I need a particular DX station that has been spotted, or has announced future activity.

- ClubLog has a "*Zone Leagues*" feature that permits analysis of: a callsign's CQ zones contacted by band ranking against other ClubLog users, globally (all users); DX Clubs registered with ClubLog, those on a particular continent, or those in an expedition registered with ClubLog. For example, as of this writing a global rank comparison shows K1ZN as being ranked 1151 globally, and 453 for the North American continent, and 17 in the Southeastern DX Club.

- ClubLog also has a "*DXCC Leagues*" feature which can make the same ranking analyses as described above, but using DXCC identities contacted by band as the criteria, rather than CQ Zones contacted.

- Using its "*Zone Chart*" feature, ClubLog can also generate CQ Zone charts showing how your call stacks up by CQ Zone. It also has a "DXCC Charts" feature which does the same for DXCC entities.

- For DX operations that have activated their logs with ClubLog, I can see my QSO's in their log as quickly as the operation uploads to ClubLog; a useful feature that has cut down the number of "insurance" QSO's that stations make to ensure that they are in fact in the log, thus reducing pile-ups on the bands.

- There is a spotting DX Cluster in place, as well. I can filter spots by band and mode, or filter spots based on my own log file in ClubLog.

- For both individual and club informational purposes, ClubLog has the ability to generate "most wanted" reports by DXCC entity. This aids clubs in funding requests for operations to remote destinations that a lot of DXers want to work.

- ClubLog developers also built in propagation prediction reporting support.

- The program also offers QSL routing suggestions for each DX operation.

- A very special feature is the *Online QSL Request Service* (OQRS) that DX operations can subscribe to. As mentioned in the chapter on QSLing, this is a program that interacts your log with the log database of the DX station. It then permits payment through PayPal™ for any fees you may need to pay for having QSL cards sent directly to you.

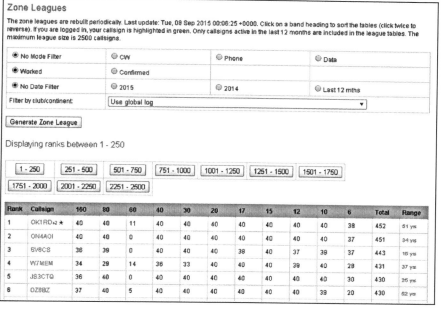

Sample of ClubLog's Zone Leagues Ranking Page
Courtesy of ClubLog & Michael, G7VJR

Note: You can filter the list of rankings using the filters shown at top of the Zone Leagues page, limiting the output to suit your purposes.

For additional information regarding ClubLog features, refer to the "ClubLog Help Desk", which is accessed by clicking on the Help button at the top of any ClubLog web page.

Additional Online Resources

DX Coffee.
DX Coffee is an online ham radio news source, a resource for keeping up with the latest happenings, and specialty social topics. The site has postings of: upcoming DX events and DXpeditions, books and resources for working DX, video clips archives, an editorial page, and propagation forecasting resources. See www.dxcoffee.com/eng

ET3AA Ethiopia
18 hours ago | 1 comment

Jan PA2P informs DxCoffee Readers: "Jan, PA2P, will visit Ethiopia for work from September 13th until September 24th 2015. He will travel around the country. When work and time permits, he will try to work from the clubstation ET3AA. ET3AA [...]

S79C Coetivy Island AF-119 [Press Release #2]
3 days ago | 0 comment

5 September 2015 Dave EI9FBB informs DxCoffee readers: S79C Press Release #2 Just over a month ago, the S79C team released information about their plans to activate brand new IOTA AF-119, Coetivy Island. With little over 2 months before [...]

ZY2QG Queimada Grande Island SA-024
4 days ago | 0 comment

Joao PU2KKE informs DxCoffee Readers: "The CabreuvaDx Group CDR Group of Brazil, will active the island of Queimada Grande, IOTA SA-024, home of the dreaded snake Jararaca Ilhoa (Botrophs Insulares). Team: PY2AE, PY2DS and PU2POP Date: [...]

Sample of DX Coffee Upcoming DXpeditions Page
Courtesy of DX Coffee, Pasqual La Gamba, IZ8IYX – K8IYX

Daily DX & Weekly DX.
An electronic publication by Bernie, W3UR, both of these subscription products provide up-to-date listings of current and upcoming

DXpeditions. Bernie digs out insights into the DX activity and gives very valuable heads-up notice of rare DX opportunities. Each edition also lists QSL routes and IOTA activity. See **www.dailydx.com**

425 DX.
The 425 DX online newsletter is a Weekly Bulletin for serious DXers. The newsletter web page is much like a dashboard: having a live search engine for callsigns, a propagation resource, news on upcoming DXpeditions, spotting history archives and other resources for serious DXers. See **www.425dxn.org**

ARRL DX Bulletin.
If you are a member of the American Radio Relay League, be sure to subscribe to the weekly DX bulletins that the league sends out.

NG3K Announced Operations.
A central resource and annual calendar listing of announced DXpeditions is available at **www.ng3k.com/Misc/adxo.html**. I rely heavily upon this web page for keeping abreast of upcoming DX operations.

WA7BNM Contest Calendar.
Somewhat similar to the DX announced operations web site, the Contest Calendar posts a calendar of contest events occurring over: the next 8 days, 12 months and perpetual contests (annually recurring). This has become a popular online site for contest postings, including: CWOps, QRP Fox Hunt, NCCC RTTY & Sprint, FISTS, IARU, SKCC, RSGB, and NAQCC contests. See: **http://www.hornucopia.com/contestcal/weeklycont.php**

Perpetual Contest Calendar 2015

| Home | 8-Day | 12-Month | Perpetual | State QSO Parties | CW | QRP |
| Log Due Dates | | Historical | Customize | | | |

This perpetual calendar shows the future dates of contests that have predictable dates. Select the desired year from the drop-down list then click on the Go button to see the schedule for any year through 2024. For a complete list of contests, use the *12-Month Calendar*, available through the link above.

Year [2015 ▼] [Go]

January 2015

SARTG New Year RTTY Contest	0800Z-1100Z, Jan 1
AGCW Happy New Year Contest	0900Z-1200Z, Jan 1
WW PMC Contest	1200Z, Jan 3 to 1200Z, Jan 4
ARRL RTTY Roundup	1800Z, Jan 3 to 2400Z, Jan 4
EUCW 160m Contest	2000Z, Jan 3 to 0700Z, Jan 4
SKCC Weekend Sprintathon	1200Z, Jan 10 to 2359Z, Jan 11
North American QSO Party, CW	1800Z, Jan 10 to 0600Z, Jan 11
NRAU-Baltic Contest, CW	0630Z-0830Z, Jan 11
NRAU-Baltic Contest, SSB	0900Z-1100Z, Jan 11
DARC 10-Meter Contest	0900Z-1059Z, Jan 11
Midwinter Contest	1000Z-1400Z, Jan 11

WA7BNM Recurring Perpetual Contest Calendar Page.
Courtesy of Bruce Horn, WA7BNM, of hornucopia.com

DX World.net.
Another very useful resource for the DXer is DX World.net. This online site provides general news of upcoming or proposed DX activities, as well as DX by continent. Most useful, in my opinion are the monthly calendars, which let you see at a glance what you might want to work. See: http://www.dx-world.net/

DX WORLD.net
FEATURED DXPEDITIONS TIMELINE

Last update: September 13, 2015

Callsign	Dates
C6ASL	~1–13
OJ0DX	~15–30
TI9/TI2HMJ	~1–9
TF/DL	~10–19
P40PX	~1–10
E6GG	~13–30
YJ0NH	~4–10
JW/OX5M	~14–20
VP5SCA	~5–12
Z21MG	~16–28
5H3DX	~5–26
T2GC	~26–30
5W0RM	~7–15
JA0JHQ/VK9C	~17–23
FO/DF1YP	~8–27
P40ER/P40ET	~10–20
3D2YA	~22–28
D67GIA	~13–23
V73YL & V73H	~22–28

Edited by MM0NDX — **SEPTEMBER** — IK8LOV for DX-WORLD.net

Monthly Dxpeditions Calendar
Courtesy of: DX World's: Wey, K8EAB and Max, IK8LOV

Chapter 6: **Documenting the Contact**

In Amateur Radio vernacular to QSL means to verify a communication or 2-way contact. In the world of "paper chasing," or working towards ham radio awards, the QSL card is a valued commodity; it serves as proof of contact.

Typical QSL card
All hams should have a supply of QSL cards.

Keeping a Log

While the Federal Communications Commission's Part 97 no longer requires Amateurs to maintain a log, for DX purposes logging is essential. Accuracy and integrity are the cornerstones of the hobby. As DXers we want to be able to obtain confirmation of successful contacts (QSOs) from the other station, and we want to be able to accurately grant confirmation to that station as well. In general practice, if we receive a QSL card from another ham, but we don't have that QSO in our log, we should not provide a card in return. Thus, this requires accurately recording the QSO when it occurs.

To Radio
K1ZN   **K 1 Z N**

Confirming our three QSO(s)

DATE	UTC	RST	MHz	2Way
12 Sept. 2014	2340	599	24.905	CW
13 Sept. 2014	0205	59	21.240	SSB
13 Sept. 2014	0306	599	18.074	CW
----------	----	---	------	---
----------	----	---	------	---
----------	----	---	------	---
----------	----	---	------	---
----------	----	---	------	---

Thank you for calling

ZL7X

Operated by JH1HRJ, JH1TXG, JE1SCJ and JA0VSH
From the   Chatham Islands, New Zealand
CQ Zone 32, ITU Zone 60,
IOTA OC-038, GL AE16QA
RIG and ANT:
 K3 + ALS600S + Inv.L (160 to 10m)
 K3 + GP (40 to 10m)
 FT897D + G5RV-jr (Digital modes)
 FT857D + 5ele Yagi (6m)

Confirmation of three QSOs with the ZL7X DXpedition

Not too many serious DXers still log with pencil and paper in a paper log book. However, this is still viable if you are not too active in the hobby, or if you really don't want to use a computer and software program. Alternatively, there are a variety of software-based logging programs available – both freeware and commercial.

QSL Logging and Contesting Software

Chasing DX requires accurate logging. If the QSO is not recorded from a DXers perspective it never happened. In fact, as I found a number of times, if the QSO's reported from either side of the contact don't align, it hampers getting a card processed, especially through the ClubLog Online QSL Request Service (OQRS) process.

So, which logging method to choose? Much of the choice depends on your operating preferences and style. Do you also contest? If so, you will want a logging program that works for contests – which requires features that are not necessary for DXing. Serious contesting requires log entry windows that can be customized for the various major contests.

Do you like "bells & whistles" such as a computer interface with your transceiver? Antenna rotator control? Automated heading information based on entered call signs? Some programs have these features. Then of course, there is the trade-off of freeware vs commercial programs, which have technical support available. Decisions; decisions! The following table shows a small sampling of the currently marketed or available logging programs – some focused on DXing – others on contesting.

Sample Logging Programs		
Program	Description	For More Info.
Logic 9	A program designed for DX logging. Good QSO entry screen. Good search capability & report generation.	www.Hosenose.com
WriteLog	A contest logging program with search & pounce features, cw decoding, call check, multipliers, packet spotting. Moderately priced.	www.writelog.com
N1MM	A free contest logging program. Supports all major contests, cw/ssb/digi, DXpedition mode possible, auto beam headings, telenet cluster support & more.	www.n1mm.hamdocs.com/
TRLog	Commercial contesting logging program with strong cw features, also supports RTTY, computer control, DVK support, vhf contests.	http://trlog.com/
DXLab	Freeware DX logging program, awards tracking, propagation support, QSL routes, callbook lookup, digital support & more.	http://dxlabsuite.com/

Online QSL Lookup Databases

eHam.net is a very popular online database and information clearinghouse containing an extensive and well used product review section, wherein logging software is listed and reviewed. **www.eham.net**. Again, some of these products are stand-alone QSO logging programs; others include utilities such as rig and rotor control, contest logging capabilities (single event logging that can be exported to special file forms for contest award entry); and programs supporting digital mode operating and rig control. Which program type you choose depends on your breadth of operating interest and style. It's easy to get carried away with all the neat features that some of these programs supply, but if you're the kind of person who hates thumbing through help files, you might be best served with one of the simpler programs. I'm a simple guy and choose a program that fits that descriptive style.

A typical QSL card exchange involves sending the DX station your card, usually including return postage and a self-addressed envelope, and then waiting for the return. Many awards programs require an actual card for credit towards the award. And, of course, we DXers love to look at the cards we receive, many are quite attractive, illustrated with images of faraway places. Cards can remind us of memorable QSOs we enjoyed.

QSL cards are an important part of our shack inventory. QSL cards can be printed directly from some of the more robust logging programs (using special card stock made for that purpose), hand printed on a U.S. Post Card – though quite impractical if you are at all active on the bands, or ordered from specialty QSL card printing companies. Again, eHam.net lists and reviews many of these printers.[viii] The card should be simple and provide easy reading of your call and include the DX's call, the date, time, frequency and mode of operation of the QSO, and a signal report to the DX station.

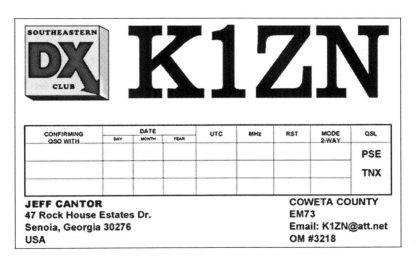

Sample QSL Card Showing Club Affiliation

QSL cards are exchanged in a number of ways. I send a card to all DX contacts made. For those *all-time-new-ones* (ATNO) DX entity or Islands on the Air (IOTA) contacts, I send the card via a direct method. Otherwise, I send the card through the outgoing ARRL QSL card bureau. Getting returns using the "buro" can take a long time, but when sending out a lot of cards, saves a substantial amount of money.

QSL Managers

Some DX hams choose not to handle their own QSL card mailings. For some, the out-of-the-way nature of their country makes receiving and sending cards troublesome – for both the hams sending requests for a card, and for them to return the cards. Others are just too busy to handle the volume of cards they receive – because of the rare nature of their location and/or because they do a lot of operating. Many DX stations use hams that volunteer to serve as their QSL manager. These methods are discussed in more detail later in the chapter.

ARRL Outgoing & Incoming QSL Bureaus

The American Radio Relay League maintains an Outgoing QSL Bureau for its members to assist in economically moving QSL cards to overseas Amateurs through Amateur Radio societies in other countries. Cost for members at the time of this writing was 10 cards for $2. Check with the ARRL for details of using the Outgoing QSL Bureau.[ix]

Incoming QSL card bureaus, sponsored locally by radio clubs, are maintained in each of the call districts in the U.S. and Canada. Cards sent by DX stations overseas and by their countries' Amateur Radio society are received by the local incoming bureaus. Volunteers sort incoming cards by recipient callsign. To receive cards, you need to maintain an account with the bureau, typically keeping them supplied with self-addressed and stamped envelopes (SASE) - check the requirements of your bureau. The number in your call determines which bureau you sign up with. My call (K1ZN) is in the first call district bureau, which is sponsored by the *Yankee Clipper Contest Club*. http://www.w1qsl.org/

The table below displays current incoming QSL bureau sponsorship by Call District.

Call District	U.S. Incoming QSL Bureaus
	Organization
1	Yankee Clipper Contest Club
2	New Jersey DX Association
3	National Capitol DX Association
4	Sterling Park Amateur Radio Club & PMB #305[x]
5	Oklahoma DX Association
6	ARRL Sixth District Incoming QSL Bureau
7	Willamette Valley DX Club
8	Great Lakes Division 8th Call Area Incoming QSL Bureau
9	Northern Illinois DX Association
0	Kansas City DX Club

QSL Direct with SASE

Direct mailing of a QSL card usually involves sending the card to the station internationally, unless the DX station maintains a QSL manager (a volunteer ham that handles QSLing for that station) in the U.S., which is not uncommon. A number of publications list QSL managers or methods to QSL, including the Amateur Radio call lookup databases on sites such as: eHam.net, QRZ.com, Buckmaster, HamCall©, Daily

DX and DXCoffee.com. These QSL method lookup source pages often provide station mailing addresses and instructions on how to obtain a card – including the postage the DX is requesting to return a card.

When I first got into the hobby, QSL'ing was a courtesy we extended to each other as hams. Today, however, the practice is to include return postage for cards mailed internationally, and stateside if a special event to offset costs of large mailings that such events generate.

When I mail a card internationally, I either include the approximate cost of postage for a single card and envelope mailed from that country to the U.S., information that I obtain by searching postal rates on the Internet, plus $2 to $3 U.S., in cash, depending on whether the card is coming from Europe or further away.

I recognize that depending on where the card is going in the world, cash might not make it safely to the DX station. To prevent postal agents of some countries from pilfering your QSL card request, for the cash it might contain, its best to avoid identifiable Amateur Radio terms (call signs etc.) on the exterior of the envelope. I sometimes find and buy foreign postage stamps and mail them to the DX. There are sources available for purchasing foreign postage to affix onto the return envelope to make it easier for the DX station to return the QSL card. For one such source see Bill Plum at **plumdx@msn.com**

PLUMDX also sells combination air mail envelopes, one smaller than the other for nesting, which is handy for placing a self-addressed and stamped return envelope inside the larger envelope with my QSL card. Then I wait for the return QSL card to arrive.

OQRS & ClubLog

More and more DX stations are turning to Internet based processes for handling QSL's. The ClubLog's *Online QSL Request Service* (OQRS), which facilitates a DXers ability to order and pay for their QSL cards has become increasingly popular. The OQRS is a component part of ClubLog, an Internet web site developed by Michael, G7VJR.[xi] Registration is free; however the site depends upon donations to cover its costs.

A valuable feature of ClubLog is that if a DX station you have worked is also using ClubLog, you can easily verify that you are in their log.

This gives you peace of mind, especially if it was an *"all-time-new-ones,"* allowing you to work additional bands and modes without worrying about that one. The OQRS feature allows you to use ClubLog as a DX station QSL card ordering mechanism, thus saving on postage to send your card to the DX station (along with cash or foreign postage stamps and a self-addressed return envelope) and waiting for a return. In turn for this savings, the DX station will probably charge a premium for ordering the card through ClubLog; payment is via PayPal. In addition to simplifying things for DX stations, this has become a good method for offsetting the cost of DXpeditions.

Log Search

This form allows you to check if you are "in the log". It only works for registered Club Log users has been heavily optimised for large and/or popular logs.

Log to search:	FW5JJ
54,327 QSOs logged between 2013-10-07 04:06Z and 2015-09-09 23:46Z	
Callsign to check:	K1ZN

[Show contacts]

Band	CW	Phone	Data
30			1
20	2		1
15	1		
12		1	
10		1	

Note: FW5JJ is also using the Club Log expedition charts

[Request QSL card]

K1ZN using OQRS Log Search to verify he is in FW5JJ's Log
A summary of contacts made is shown and a QSL can be requested.
(Courtesy Club Log - (c) Michael Wells, G7VJR)

From the perspective of a DXer, once you establish an account by registering your call and uploading your log, ClubLog has the capability of generating reports and analyses of the DXCC activity recorded in your log. These reports can be run on an ongoing basis to analyze: DX activity per year, DX activity per band and mode, and for the DX entities still needed.

ClubLog is a wonderful resource and tool for DXers. As a DX shack tool, ClubLog has the capability of searching the logs of DX stations you have worked (providing the DXer chooses to upload it), and propagation prediction tools to aid you on the bands. It maintains a list of most wanted entities based on an aggregate analysis of all ClubLog user logs, letting DXpedition planners know what entities will attract the most interest.

A real time DX Internet spotting tool is a must in any DX shack and ClubLog provides this capability specifically tailored to your log, identifying and presenting needed band slots, saving you the chore of wading through spots to find ones you need. I suggest that early-on in your DX pursuits you take the time to become familiar with ClubLog's capabilities and use this tool. ClubLog is also discussed in Chapter 5.

LogBook of the World (LOTW)

LogBook of the World is a software tool developed by the American Radio Relay League in support of its DXCC award program. The LOTW program requires that ARRL members apply for a software "key" (called a Callsign Certificate) to initialize the LOTW software on your computer. Then by uploading your QSL log to LOTW, your log data can be matched against other DXers logs and QSOs, confirming those where the contact information (time, mode, frequency, call) match. Periodically, you can apply those confirmed credits against your DXCC awards, and also towards your WAS (Worked All States), VUCC (VHF/UHF Century Club) and WPX (CallSign Prefix) awards if you are also working on those programs.

At the time of this writing, the ARRL does not have award program credits reciprocity with the *Radio Society of Great Britain* for their *Islands on the Air* (IOTA) program or with any other world Amateur Radio society for any other awards program (e.g., Worked All Zones; Worked All Continents).

E-QSLs

While the E-QSL program exists and is economical to use, it is not accepted in the DX world as proof of a 2-way QSO.

DXCC & IOTA Card Checking

Both the DXCC and IOTA awards programs use a field checking process to expedite the validation process for submission of cards. In the DXCC program, a list of approved field checkers can be found on the DXCC web page. Approved card checkers generally appear at major hamfests, and will check your cards on a walk-in basis. However, be sure to bring your web-initiated award application, a stamped envelope addressed to the DXCC Desk at the ARRL, and the award's application fee. Similarly, on the RSGB-IOTA web page, a checkpoint is assigned to you once you initiate a first application for the IOTA award.

Other awards, including *Worked All Zones*, also use field checkers. For details, check the relevant awards web page or printed information concerning submitting confirmation for individual awards.

Chapter 7: Awards and Paper Chasing

Chasing distant and foreign radio stations as a hobby is very much like golf – it is a personal challenge to contact as many as possible, potentially qualifying for various awards. In ham radio, this is termed "paper chasing," – chasing an award: certificate or plaque. The pursuit of awards gives a sense of purpose to ham radio, giving the feeling of being on the hunt as needed contacts are sought and made, followed by a sense of satisfaction when an award is actually earned. There are a number of major awards that hams can strive for. Here are some of the more popular ones.

DXCC Award

The American Radio Relay League sponsors the *DX Century Club* (DXCC) Awards program.[xii] The basic award is for presenting confirmation (QSL cards) of having made two-way contact with licensed Amateur Radio operators in 100 of the currently possible 340 world's countries or entities. These contacts can be on any of the HF bands and in any mode (Phone, CW or Digital). There are 19 different ARRL awards certificates available, plus the Honor Roll and Challenge awards. Once a ham has achieved 100 countries/entities on each of the basic 80-40-20-15-10M bands, then the 5-Band DXCC award can be applied for.

After achieving the 5-Band award, the next step is to obtain awards for the 17M, 12M, and 30 Meter bands and finally for 160M and 6M. Endorsements for 100 countries/entities are available for each mode as well – CW, Phone, Digital and Satellite. A VHF/UHF version of the Century Club award is also available for 2m, 440cm and above.

The ultimate goal in the program is to achieve Honor Roll and finally to work all current entities – which at the time of this writing totals 340. Honor Roll requires working at least 331 of the current entities. Deleted entities do not count for awards. A deleted entity is a country or political entity that ceases to exist or be viable for DX purposes. With the breakup of Czechoslovakia, for example, this country ceased to be an entity.

Finally, the *DXCC Challenge Award* is available for working at least 1,000 DXCC band-entities on any of the bands. This award is endorsable in increments of 500 entities (any band – any mode) to keep you moving along!

There are several ways to apply for DXCC awards. The ARRL's online *LogBook of the World* (LoTW) is the easiest, as no QSL cards need to be submitted with an application. However, DX stations you have worked need to have uploaded their logs showing your QSOs for this to work. You first have to establish a LoTW account with the ARRL. If you are a U.S. Amateur Radio operator, you need to join the ARRL to do this. Once your account has been established you can upload your log (most every logging program supports this capability). When both the DX station's QSOs and your QSOs match in LoTW they will be displayed as confirmed contacts. Upon reaching the century mark (100 entities – any band – any mode) you may make application through the LoTW program for your DXCC award certificate.

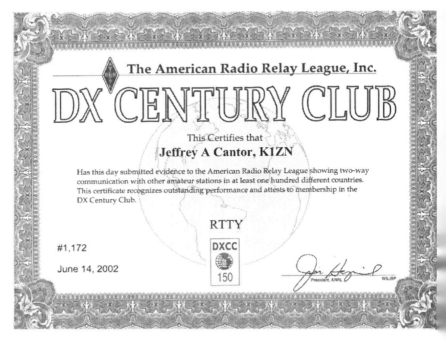

DXCC Award Certificate

Most likely, you will also have some QSL cards in hand which will need to be verified, having sent for them directly. And not all worldwide DX stations use, or care to use the ARRL LoTW process. Consequently, you might need to do both an online LoTW credit submission and a paper submission for the QSL cards.

Paper submission of cards can take two forms. The method easiest for most hams is to fill out the ARRL's on-line DXCC application form available at **http://www.arrl.org/online-dxcc-application**, and take that paper application and the QSL cards listed on that application to an authorized ARRL DXCC card checker. These card checkers are listed by zip code location on the ARRL DXCC web site. They are usually affiliated with local DXCC clubs. If you cannot locate one within reasonable travel distance of your QTH, then the second of the two paper application methods is to send the application you created on the web with the accompanying QSL cards directly to the ARRL DXCC Desk, at 225 Main Street, Newington, CT 06111. You will also need to include payment for the application and number of cards submitted, plus postage for return.

The above described process is the same for all additional DXCC awards and endorsements you may apply for.

There are other organizations that sponsor awards for working all or most of the world's countries or international entities. I will touch on some of these later in this chapter.

Worked all States Award

A usual initial goal in the paper chasing game is to earn the *Worked all States* award, which has as its objective the making of two-way radio contacts with all 50 states. The ARRL recognizes this accomplishment with a nice certificate upon QSL card or LoTW proof of having made the required number of contacts on any combination of HF bands and modes. As you continue to make contacts, the certificate is then endorsable for phone, CW, and digital modes when all 50 of the states are contacted in said modes. Upon achieving contact with all 50 states, using any mode, on each of the 5 HF bands (80-40-20-15-10M), a five-band plaque can be applied for. Endorsements for the 12/17/30M bands can also be earned. A top-band endorsement for working all 50 states on 160M is tough to do, but is achievable. The WAS rules can be found at **http://www.arrl.org/was**

Five-Band Worked All States Award

VHF/UHF Awards

The ARRL also sponsors the *VHF/UHF Century Club* (VUCC) awards programs. This awards program is based upon the maidenhead grid square system. The basic award requires making contacts in 100 grid squares on 50 MHz and 10 grid squares on 144 MHz. On the UHF bands, 222 MHz and 432 MHz, only 50 grid squares are required for the basic awards. I enjoy the challenge of chasing grids on 50 MHz and have qualified for this award.

Islands on the Air Program & Award

A program I find fascinating is chasing the many islands of the world. The Radio Society of Great Britain sponsors the *Islands on the Air Program* (IOTA). This is a program of some 22 different certificates and two plaques being awarded for acquiring two-way radio contact with DX operators located on over 1200 islands groups worldwide. The basic award is IOTA 100, which is granted for achieving confirmation of your first 100. Then, in increments of 100 you work towards IOTA 200 etc., up to IOTA 1100. There are also certificates for achieving a maximum score on each of the seven continents, British Isles, West Indies, Arctic Islands, and finally a *World Diploma*.

IOTA QSL Card

RSGB-IOTA has an excellent web page and online application system which permits you to enter and monitor your QSL confirmations as you amass them, and for then submitting electronically into the RSGB database, followed by sending the QSL cards to your assigned checkpoint for validation.

IOTA 500 Islands of the World

IOTA cards must have the specific island of operation imprinted on them as well as the IOTA number, when possible. The exact island name is the most important feature of an IOTA QSL card.
See: http://www.rsgbiota.org/

RSGB IOTA Home Page
Courtesy of Radio Society of Great Britain

Worked all Continents Award (WAC)

The International Amateur Radio Union (IARU) is the sponsor of the *Worked All Continents Award* and the *Five-Band WAC* award. To qualify you must work all six continents from at least one country or entity. You must have a QSL card verifying the necessary two-way international contacts. Application is made through an IARU member Amateur Radio Society such as ARRL. See:
http://www.iaru.org/worked-all-continents-award.html

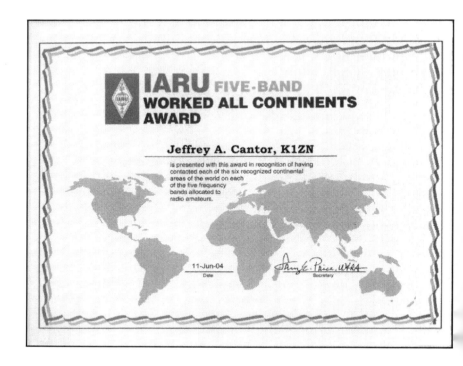

Worked All Zones Awards

The Radio Society of Great Britain administers the ITU's *Worked all Zones* program. The ITU *Worked all Zones* program requires QSL verification for having worked 70 of the 75 International Telecommunications Union zones. Five-band and WARC bands (17M-12M-30M) endorsements are also available. For more information, see: http://rsgb.org/main/operating/amateur-radio-awards/rsgb-hf-awards/worked-itu-zones/

CQ Worked All Zones, WAZ

CQ Magazine instigated an awards program that parallels various ARRL, RSGB and IARU programs in many respects. Awards for DX entities, *Worked all States*, and *Worked all Zones* are offered. The *Worked all Zones* program requires proof of two-way radio contact with all 40 CQ zones – which are different from the IARU ITU zones, as shown by the zone maps in chapter 3. The *WPX Award* is for working international prefixes, both common and uncommon prefixes. See: http://www.cq-amateur-radio.com/cq_awards/index_cq_awards.html

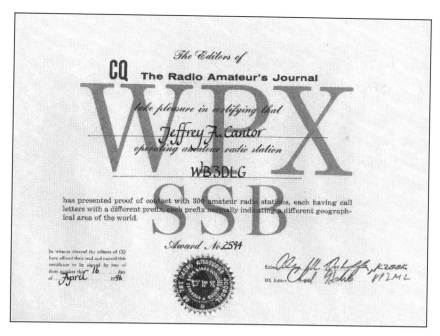

CQ's 300 International Prefixes and Worked All Zones Awards

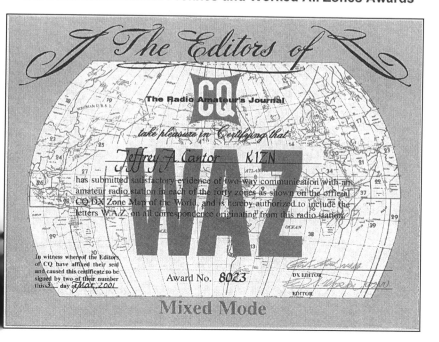

Various Countries Also Sponsor Awards

Ireland's radio society, the Irish Radio Transmitters Society sponsors the *WAI Counties Award*. Radio Amateurs of Canada (RAC) sponsors several awards including an award for working all Canadian provinces. The *Helvetia Contest* is an annual event to work all Swiss cantons. In addition, there are numerous paper chasing events being sponsored by various organizations – probably at least one a weekend. The Deutschland Radio Club (Germany) sponsors the *Worked All Europe Award* in several levels for working European country stations. See the table of international DX awards at the end of this chapter.

Deutschland Radio Club (Germany) Worked All Europe Award

Stateside Operating Awards

DX can mean different things to different ham radio operators. I lik traditional international station DXing. I also like chasing island When not doing either of those, I like chasing U.S. counties.

U.S. county hunting is a very popular activity on the HF bands. Daily, on 14.336MHz, 7.188MHz, and 17M frequencies as announced, you will find county hunters aspiring to the USACA awards.

The *United States of America County Award* program (USACA) was started by CQ Magazine, and continues being popular. There are officially 3,077 U.S. counties. A certificate is issued after achieving the first 500 counties, with 500 county endorsements up to the ultimate top plaque. Rather than sending and receiving QSL cards for each county, the program uses a mobile reply card which conveniently records up to 8 county contacts per mobile station worked. Frequently mobile ham radio operators broadcast while moving from county to county, thereby putting those counties on the air, some of which may have very few or no hams in residence.

Confirming QSO:
FROM ARS:_____ TO ARS: **K1ZN**
Mode CW☐ SSB☐ FM☐ MOBILE☐ PORTABLE☐ FIXED☐
MARAC #_____ USCA #_____ 2nd Time#_____ 3rd Time#_____

Date	UTC	MHz	RS(T)	State	County

Signature: _____   www.cheapqsls.com

Mobile Reply Card for recording multiple counties

Alternatively, you might simply work a U.S. station and ask for the county they are broadcasting from. More information at: http://www.cq-amateur-radio.com/cq_awards/cq_usa_ca_awards/cq_usa_ca_awards.html

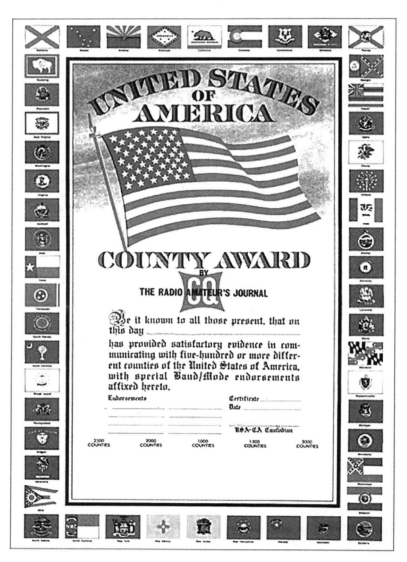

CQ Worked US Counties Award
Courtesy of CQ Magazine

Individual state awards for worked all counties.
Some clubs and organizations sponsor awards within their states for worked all counties. The Candlewood Amateur Radio Club of Danbury Connecticut did so, giving me the opportunity of earning an award for working all eight Connecticut counties.

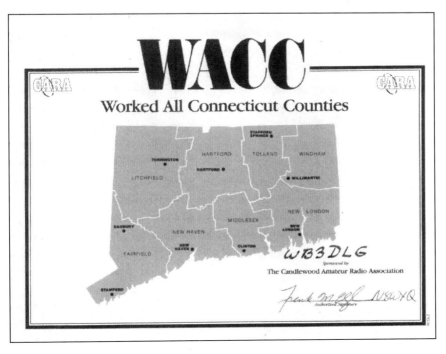

Award for Working all Eight Counties in Connecticut

Program	Description & Web Link
Radio Amateurs of Canada	RAC sponsors: a Canadaward plus a 5-Band version, Provincial Capitals Award, Trans Canada Award, and Worked all RAC Awards. https://www.rac.ca/rac-awards/
Islands of Croatia Awards	Several awards – checkpoints & qualifying islands. http://www.inet.hr/9a6aa/IOCA_IOTA_LH/detail.htm
Czech Amateur Radio Club Awards Program	CRC issues several awards for working Czech ham stations. http://www.crk.cz/ENG/AWARDE
(Deutch) German Amateur Radio Club	DARC sponsors "diplomas" for DX & Worked All Europe. See: http://www.darc.de/referate/dx/diplome/dld/en/
European World Wide Award	Issued by the Radio Club of the Council of Europe. Basic award is for confirmed contacts with 200 different countries of EWWA list. http://tp2ce.eu/en/2012-03-25-17-30-09/2012-03-25-17-30-45/award.html
Irish Transmitters Society	WAI Worked all Counties Award http://www.irts.ie/cgi/wai15.cgi
Israeli Awards	Holy Land Award – contact 100 different areas of 13 regions of Israel. http://www.dxawards.com/DXAwardDir
Japan Amateur Radio League (JARL) Awards Program	Numerous awards for contacts with Japanese Amateur Radio operators, including a 90th JARL Anniversary award. http://www.jarl.org/English/4_Library/A-4-2_Awards/Award_Main.htm
South African Radio League Awards	SARL sponsors 3 awards, All Africa Award; Worked all ZS; and Top Band Award. http://www.sarl.org.za/public/awards/awards.asp
Wireless Institute of Australia	Multitude of awards including a series of DXCC awards. http://www.wia.org.au/members/awards/about/
Indonesian Islands Hunters Group	Nusantara award for contacting at least 10 different islands of Indonesia. http://www.nusantaraaward.com/

Chapter 8: DXing on the Top and Magic Bands

Top Band DXing

Are you a late night person? Do you have lots of patience for listening, especially through abundant QRM static noise crashes? If yes, then 160M DXing could be for you. Most DXers on 160M have been on the HF bands for some time, earned DXCC on those bands and are looking for a new challenge.

The top band presents unique challenges in pulling out signals from the noise usually present on this band. Additionally, 160M becomes active at night and into the very early morning, so staying up late and getting up early is a must. The band becomes less noisy in the winter months, making that an ideal time for working the band.

160M Propagation Characteristics

Much of the uniqueness of 160M is due to its propagation characteristics. Higher atmospheric absorption levels result in weaker RF signals. Signals peak at sunset local time on the western side of a given signal path and at local sunrise on its eastern side. Openings even in the winter months, however, can be short. Signal fading is a known top band characteristic that has to be accepted. Here are some best operating practices and times for trying the band.

Listening Effectively

Atmospheric QRN, static crashes and the like make working 160M DX extremely challenging, requiring very careful listening. Effective listening on this band requires a good set of noise-cancelling headphones and your unfettered attention. Use of the radio's filters and DSP settings is essential for minimizing the effects of the noise. Additionally, multiple antennas, especially dedicated receive antennas are a good idea.

Listening Antennas

Beverages.
Those who seriously DX on 160M have found that dedicated receive antennas are a must. Designs that minimize noise is essential, these include: low mounted dipoles, receiving loops and Beverages. These usually require a wire cut for the band and be suspended close to the ground. The Beverage antenna is named after the chap who first pioneered the design. Plans for Beverage antennas can easily be found on the Internet, some featuring various refinements in construction and mounting techniques.

Receiving Loops.
The K9AY design is another popular form of listening antenna. This design is good for those with limited space on their property that prevent using a Beverage design.

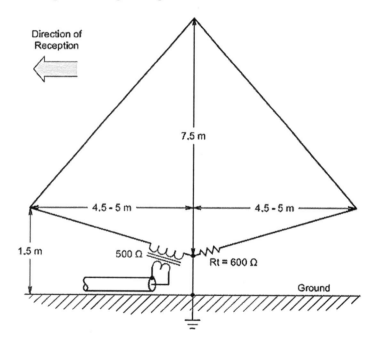

K9AY Loop, Notice matching transformer and terminating resistor
Courtesy: http://www.ok7k.estranky.cz/clanky/rx-antennas/k9ay-loop/

Another space limited design is called a flag receiving antenna, while more complex in construction; it is removed from ground, and therefore quieter.

In his book, *"DXing on the edge"* K1ZM advises that short vertical arrays are gaining in popularity with those that do not have enough space for erecting Beverages or other large receiving antennas. Short vertical arrays can be designed as 3 verticals placed in triangle form, or 4 verticals in a square. They are also available commercially.

There are other forms of wire antenna designs, as well. The basic principle is to outfit your station with an antenna that minimizes band noise, enabling you to copy weak and fading signals.

Transmit Antennas

On the transmit side, I have successfully used a loop antenna resonant on 160M. This is not a small antenna; one needs over 540 ft. of wire length to resonate on this band. My loop is ladder line fed and terminates in a high power balun. I recently constructed an inverted-L antenna, mounted to my tower. The radiator is 126.5 ft., with the vertical portion at 53 ft. At the top, the radiator is bent at a right angle and horizontally strung, supported by an adjacent tree. Completing the design, three 135 ft. radials, raised to 8 ft. are attached to the ground side of an SO-239 connector.

Top band transmit antennas also include tower loading, vertical phased arrays, dipoles, four squares, etc.

For some good reading on both transmit and receive antennas, I recommend Jeff Briggs' (K1ZM) book, *"DXing on the Edge,"* published by the ARRL.

On 160 QRO is a Must

Serious DXers have learned that QRO (increased or high power up to the maximum allowed by law) on 160M is almost a must. The band's propagation characteristics, QRN and fading all make for a very difficult contact. Good antennas are essential; but even so, using high power will often be necessary for completing a contact before signals disappear. And, if there are competing stations, you may need the extra power to be heard.

K1ZM's Recommendations

Jeff, K1ZM makes recommendations in his book for successfully working 160M DX pileups that are worth repeating. Some of these suggestions were made earlier in the operating practices chapter, but they are worth repeating.

1. Listen first, then call after you are sure of the DX stations listening frequency. If they are working another station don't call on top of that QSO. Listen first! Use a spotting cluster to try and get information on whether the DX is operating simplex or split – and where they are listening.
2. Listen carefully to be sure that the DX has not changed their listening frequency (QSX). Very often on 160M a station will be experiencing so much QRN that they try to find a better spot to be able to hear incoming signals.
3. Don't step on another operator trying to get through to the DX. If the DX is calling another station or trying to clarify their call, don't try to step on top of them. In all probability the DX station won't give up on the other station in favor of you anyway.
4. Know your rig and how to use it. Know how to zero beat a signal. Know how to use your filters effectively. Know how to adjust transmit audio, compression, etc. Don't put out a bad signal which not only affects your ability to make the contact, but disrupts all others on frequency.
5. Focus on building good receive antenna capabilities, first. Then work on improving your transmit capabilities.
6. Transmit message clarity is essential on 160M. When operating CW that means: sending clean CW – with correct spacing, sending full calls only, and matching the DX station's sending speed.
7. Ensure good contacts by sending the DX station's call to ensure that it is, in fact, the correct call.
8. Listen first!

Six Meters - the Other End of the Spectrum

The magic band, 6 Meters is equally as fascinating. As a VHF band, 6 Meters is open to Technician licensees, as such, this can be the first band a new ham can use to start DXing. This band can also be frustrating, however, in that openings are very sporadic with long periods of time lapsing between DX opportunities.

I have been DXing on this band for over four decades. As a testament to how challenging this band can be, I'm sorry to say I am still working towards DXCC on 6M. To be productive on this band one must both monitor the band continuously and watch for spots on the cluster. Persistence is a must for earning the coveted DXCC on 6 Meters award.

Six Meter Propagation Characteristics

When solar activity increases ionization levels in the upper atmosphere such as during sunspot maximums, the 6 Meter band can act much like an HF band. I recall during the 2005 sunspot peak, I was able to work throughout Central and South America.

In general, Northern Hemisphere activity sporadic E propagation peaks from May through early August, which facilitates long distance communication up to about 1600 miles for single-hop skip. In the Southern Hemisphere, intercontinental multiple-hop sporadic E propagation is possible up to 6200 miles or so, most commonly occurring from November through early February. Single-hop skip happens as summer progresses. Sometimes we see double-hop skip as well. And occasionally, TEP (transequatorial propagation) is experienced, as discussed in chapter 4.

Equipment Needs

Most all modern transceivers contain 6M capability. When the band is open, 100 Watts is sufficient for making a good contact. However, a linear amplifier will move you to the head of the line in getting through to a rare DX station. And most modern linears are capable of operating on 6M, as well.

Antenna requirements include a beam, if possible, or a dipole cut for 6M and strung up as high as possible, oriented to the direction that you are wishing to DX. I began with a dipole, and worked into South America and the Caribbean without too much difficulty. Then I graduated to a 3-element beam and worked well into Europe from my QTH in New England.

Awards & Programs

Six Meter DXing offers some interesting awards opportunities. These include the traditional worked all states programs, DXCC – 6 Meters, and also VUCC (VHF/UHF Century Club) grid squares chasing.

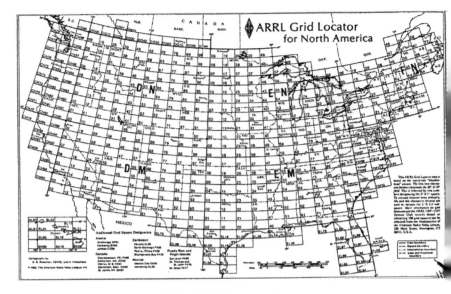

U.S. Grid Square Map
Courtesy: ARRL, Amateur Radio Relay League

I enjoy grid square chasing in addition to DXCC. Work a new DXCC entity and you also work a new grid square! And yes, the IOTA program also has a VHF category. Enjoy the challenges and rewards of working 160M and 6M, two very special bands.

Chapter 9: Unique Challenges of QRP

Most DXers, me included, began chasing DX with a basic 100 Watt rig. In my case, my first HF rig was a Radio Shack mobile size 10 Meter rig that was rated for less than 100W – I believe it produced about 30 watts. I was pleasantly surprised when I was able to make contacts with stations in South America and Western Europe very easily with only a dipole antenna. That said, the challenge of QRP – defined as working with 5 Watts or less, can be quite exciting!

Integrating QRP into Other Activities

Within my circle of Amateur Radio friends, QRP appears most enticing to those that enjoy outdoor activities as much as, or more than ham radio and DXing, such as my friend Barrett, KE4R. With smaller and more compact gear, working QRP can be combined with: camping, hiking, bicycling or even water sports and aviation! Setting up an antenna in a tree, or using a Budipole (more about this in a bit) at a campsite is quite easy. Sitting around a warm camp fire in the evening working DX on 40 Meters can be a relaxing and rewarding way to end the day.

QRP Transceivers – Factory Made or Build Your Own

A transceiver for operating QRP can be your regular HF/VHF transceiver operated with the RF drive power cranked down to 5W or below, or it can be a dedicated low power rig, many of which are sold in kit form. An example is the Elecraft KX3™, available both as a kit or a completed product.

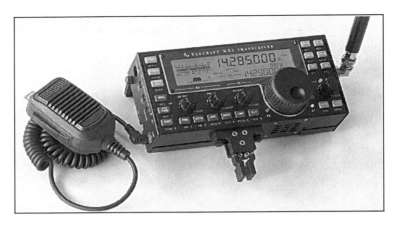

Elecraft KX3™ QRP HF Transceiver
Courtesy Elecraft Corporation

The Ten-Tec QRP Rebel™, below, is another rig specifically designed for low power portable work. This radio features open source software code, enabling the operator to alter the programming to suit one's needs.

Ten-Tec Rebel™ QRP Transceiver
Courtesy of RKR Designs, LLC

There are numerous other kits and other factory built rigs available, plus plans and schematics to build your own QRP rigs from scratch are periodically published in QST Magazine. If you are inclined to do-it-yourself, take a look at the archives of QST on the ARRL web page to research sample plans. Another alternative for do-it-yourselfers is the

American QRP Club, an organization dedicated to various aspects of QRP operation **http://www.amqrp.org/** The club supports: kit building, a forum, a magazine dedicated to homebrewing, and is an excellent resource for low power enthusiasts.

Portable Antennas

Many hams on the move, either vacationing or on a combination hike / ham radio outing will take along the fixings for a wire antenna for the band or bands they plan to operate on. While a little more bulky than a simple wire antenna, the portable multiband Budipole™ antenna has become very popular for this kind of outdoor operation. W3FF, who goes by the nick-name Budd, developed early versions of this antenna in 2000 and it has undergone refinement ever since. The antenna can be homebrewed, or purchased from **www.buddipole.com**. To locate W3FF's homebrewing plans, perform an Internet search for "homebrew buddipole." Budd, W3FF, is shown here operating mountain top portable with a Buddipole at Gray Butte Summit, California.

W3FF Operating Mountain Top Portable
Courtesy of Buddipole, Inc.

Signing QRP

Operating with low power is not all that different from any other kind of operating. However, to alert other stations of your operating condition, it is customary to sign as "/QRP," thus if I was calling CQ, I would transmit, CQ CQ CQ de K1ZN/QRP, K1ZN/QRP." Upon hearing this, most stations will typically grant preference over full power stations in acknowledging your call and making the contact.

QRP Organizations & Groups

For those getting involved with QRP operation, there are several groups and clubs which offer a variety of support. *QRP Amateur Radio Club International* is one such organization, which produces a magazine, sponsors multiple operating awards, publishes technical articles, provides kit building information, etc. The web site is: http://www.qrparci.org/

Another, *QRP Club* is a web-based forum for QRP enthusiasts. See: http://www.qrpclub.org/qrp/

The *North American QRP CW Club* is another organization promoting and supporting QRP operation. This organization sponsors monthly: sprints, CW nets, awards, assistance, etc. See: http://www.naqcc.info/

The ARRL has a dedicated page on their web site for QRP operation. Their "*QRP – Low Power Operating*" page has references and links to: QRP articles, various state QRP clubs, and other web sites dedicated to this aspect of the DX hobby. See: http://www.arrl.org/qrp-low-power-operating

Where to find QRP Stations

Over time QRP stations have established operating practices whereby they frequent certain portions of the bands. Much like island hunting or RTTY, there are published frequencies where they most likely can be found. The table below lists frequencies by band where QRP activity is likely to found.

QRP Activity Frequencies		
BAND	CW (MHz)	PHONE (MHz)
160	1.810 1.843	1.910
80	3.560	3.985 3.690
40	7.030 7.122	7.090 7.285
30	10.106 10.116	none
20	14.060	14.285
17	18.096	18.130
15	21.060	21.285 21.385
12	24.906	24.950
10	28.060	28.345
6	50.096	50.185

Operating Activities

Several of the previously mentioned organizations sponsor QRP operating activities.

QRP ARCI sponsors:
- Hoot Owl Sprint – a CW only evening activity
- Spring QSO Party
- Fireside Sprint – SSB only
- Pet Rock Sprint – CW only / crystal controlled rigs

North American QRP CW Club sponsors:
- Monthly sprints
- Monthly challenges

Awards & Certificates – QRP

Several of these organizations also sponsor QRP operating awards.

QRP ARCI sponsors:
- 1000 Miles per Watt Award
- County Hunters
- QRP Worked all States
- QRP Worked all Continents
- QRP Grid Squares

The ARRL offers:
- DXCC – QRP

Tips for QRP DXing or Contesting

❖ Sometimes it's best to let the big guns make their contacts first and then come in later – maybe the second day or so on multiday events.
❖ Listen for clear spots within the split range
❖ Concentrate on building the best antenna possible
❖ Take the helm, call CQ in a clear band spot and let them find you

Chapter 10: Considerations for Going QRO

At the other end of the RF power operating spectrum is QRO, operating with enhanced power. Any gathering of DX hams will inevitably bring on a discussion of power. You worked a new one? On what band? How much power were you running? Which amp do you have? It is interesting to note just how many of us tend to collect amplifiers; especially since amplifiers are often more expensive than a new transceiver. What does an amp do for us that makes it so precious and sought after?

Increased Power Operation

We all want to have a "big gun" signal. The signal heard first and loudest! Yes, there are a number of ways to be heard first. One is to transmit where the DX station is listening at the moment. More often than not this is just plain luck in timing your call.

As mentioned earlier, maximize your effective radiated power by using a low loss transmission line cut as short as possible. Employ an antenna system that is as effective as possible – a beam with sufficient elements to ensure maximum gain on the band(s) you wish to operate. Place directional antennas up in the air as high as possible (65 ~ 70 ft. optimally for HF). Take advantage of the real estate geography that you have – lakes or other bodies of water reflect RF nicely.

Be mindful of the fact that if you can't hear the DX, you can't work them. So, consider a separate listening only antenna, the Beverage discussed in the prior chapter is a good choice. Especially when working the low bands (40-80-160), a listening antenna can be very handy.

OK, so you have the best antenna farm you can muster up and a decent transceiver with optimum filtering for tuning in the signal. You have mastered the art of working split so you can identify where the DX is likely to be listening next, and you can clearly send your call. What's next?

Linear Amplifiers

Many of us eventually reach a point in our DX career where we decide that "barefoot" is not enough. Barefoot is the term for operating only with the power capacity of our rigs – usually 100W for most modern transceivers. If you still have one of the older "boat anchors" like a Swan 500 or a Collins rig, those that employ vacuum tube finals, your barefoot is probably 350 ~ 400W or more. For us 100W folks, though, we can turn to a linear amplifier to give our signals that extra boost. Well then, which linear amp? How much is enough? The FCC tells us that for the majority of our bands the upper limit is 1500W PEP, which gives us the upper bound.

In our world, linear amplifiers are classified two ways. First, by power output, and then by type of power amplifier (tube or solid state device) being employed to boost the drive signal from the transceiver.

Tube amplifiers, the older of the two, have been around a very long time and are still manufactured today. There are numerous tube version amplifiers on the resale market, which might be a good place to start your search. Common tubes include the 572B and 811A and the 3-500Z triodes. Amplifiers that use the 572B & 811A tubes usually use them in pairs for about 600W of output, or in a few cases four tubes, for producing around 1000W output.

811A Power Triode Tube

Typical 3-500Z equipped amplifiers can get 600W or more from a single tube or 1300W using a pair of these tubes. These are called "near" legal limit amplifiers.

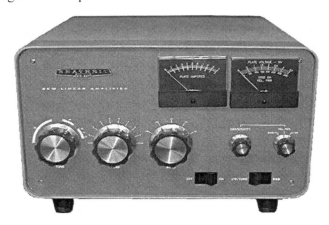

Heath SB-220™ – Uses a pair of 3-500Z Tubes
Tube Amplifiers are still Popular Today

Tube amplifiers must be manually tuned prior to use, and every time a significant frequency change is made, or the band or mode of operation is changed, which some may consider a disadvantage. There are very few exceptions to this, which can be found in the very high price range amplifiers.

The second classification of amplifier is the solid state amplifier. The solid state linear amplifier is an automatic tuning amplifier, and in a few cases can be controlled by the transceiver to automate band and mode switching.

Over the last several years a number of solid state amplifiers have appeared on the market, ranging from medium power (600~800W) to full legal limit (1500W). Amplification is accomplished using MOSFET transistors having an average power rating of about 300W. My linear amplifier, the Elecraft KPA500, uses a pair of these and is rated at 500W-plus. I am able to obtain between 600~650W of output power using a transceiver drive power of about 25~30W. Drive power refers to the transceiver RF power output setting used to drive the linear amplifier.

Elecraft KPA500™ Solid State Amplifier
Courtesy of Elecraft Corporation

There are other manufacturers selling 600W or more solid state linears as well. If you are contemplating purchasing your first linear amp, check out the QTH.com classifieds for a multitude of amps on the resale market.

Other QRO Station Considerations

Once you have become comfortable DXing barefoot, a future consideration for enhancing your station's effectiveness might very well be going QRO. But before making that decision, be sure your antenna system is as effective as it can be. Dollar-for-dollar, money spent on improving your antenna system generally produces more bang for the buck.

Also be sure to consider the impact the addition of a linear amplifier has on the rest of your station setup. Make sure your antenna and feedline are rated for the higher power levels you intend to use. Pay particular attention to the power ratings on any traps, matching transformers or baluns you might have in your system. If you are using a tuner, it also needs to be rated to handle the higher power levels. And don't forget the amount of AC power you need in the shack. Full limit amplifiers may well require a 220 VAC outlet or a dedicated 115 VAC 20 amp power outlet.

Chapter 11: DX Clubs and Organizations

To help DXer's find others close by to associate with, and provide help and mentoring when needed, I have collected as many local and regional clubs as I could find at the time of this writing. This list is by no means complete. I'm sure there are plenty of other good clubs out there that you might find interesting. To find out more about any of these clubs, or to contact them, search the Internet for the club name.

Some Local & Regional DX Clubs	
Central Arizona DX Association	Oklahoma DX Association
Central Oregon DX Club	Potomac Valley Radio Club
Central Texas DX & Contest Club	Redwood Empire DX Association
Fort Wayne DX Association	San Diego DX Club
Kansas City DX Club	Southeastern DX Club
Lone Star DX Association	Southern California DX Club
National Capitol DX Association	South West Ohio DX Association
New Jersey DX Association	Spokane DX Association
Northern California DX Club	Texas DX Society
North Alabama DX DX Club	Twin cities DX Association
North Florida DX Association	Yankee Clipper DX Club
Northern Illinois DX Association	Western Pennsylvania DX Association
Northern Ohio DX Association	Western Washington DX Club
	Willamette Valley DX Club

Joining a DX Club

Consider joining a local DX club, as they can be of invaluable assistance to those getting started into DXing. Clubs frequently present programs of interest such as the "ins-and-outs" of gaining government permission to go to an island or entity for a DXpedition, and presentations about qualifying and applying for awards programs. Club meetings provide useful insights that you might not get any other way.

Like many clubs, the *Southeast DX Club* (SEDXC), which I belong to, often feature presentations at meetings. Recently we were fortunate to have Bob, K4UEE of K1N fame, make a presentation on what it took to get onto Navassa Island, an uninhabited Caribbean island, from a governmental administrative perspective. We also had Dan, W4TKS present on the Islands on the Air (IOTA) Awards program.

The club's members are always ready to assist each other with antenna projects or with sharing of equipment expertise. Some clubs also work to raise money to donate to worthy DXpedition projects, especially those in the top most wanted categories.

Fundraising for Dxpeditions

In addition to corporate and club donations, DXpedition funding can also come from organizations especially created to promote DX operations while raising funds to help finance them.

Northern California DX Foundation
This organization was established to support and provide funding for worthwhile Amateur Radio and scientific projects, worldwide. It is entirely privately funded, and most serious DXers recognize NCDXF's value to the hobby and participate accordingly.

The NCDXF web site provides links to numerous bulletins and DX resources: QSL routing information, callsign lookup databases, propagation resources, licensing information for DXers seeking to operate abroad, links to DX clubs and organizations, and to a network of beacons for assessing propagation.

Membership applications can be initiated via their web page. NCDXF has provided funding support for numerous DXpeditions that was essential to their viability. Projects included the recent K1N DXpedition to Navassa Island, one of the most expensive ventures undertaken by the Amateur Radio community in years. See:
http://www.ncdxf.org/pages/dxresources.html

Supporting organizations often receive recognition on DXpedition's QSL cards
Note logos of the organizations providing support to this DXpedition

International DX Association

One of the first organizations that I achieved life membership in is the International DX Association, INDEXA. Formed to promote international good-will via Amateur Radio, INDEXA has become one of the preeminent organizations sponsoring DXpeditions worldwide. The INDEXA Officers and Board of Directors recently announced their intent to encourage DXpeditions to provide humanitarian aid and services that benefit individuals and communities they visit on DXpeditions. See http://www.indexa.org

YASME Foundation

The YASME Foundation is a not-for-profit corporation organized to conduct scientific and educational projects related to Amateur Radio, including DXing and the introduction and promotion of Amateur Radio in developing countries. For more information see The YASME book, "YASME--The Danny Weil and Colvin Radio Expeditions" by James D. Cain, K1TN. See: http://www.yasme.org/

Major DX Get-togethers

International DX Convention

The International DX Convention is one of the premier gatherings for DXers of all kinds. This annual three-day event, held in Visalia

California, is an opportunity for Amateur Radio enthusiasts with a focus on DXing to come together from all over the world to share their interests and stories. In April of 2015, over 800 hams gathered for a 3-day program. The first day is dedicated to DX University and Contest University programs, where those new to DX or contesting can hear about the fun, challenges and best practices from veteran DXers. Forums on various topics and displays of new equipment follow on Saturday and Sunday morning. This is a must event for DXers and contesters alike.

ARRL Divisional Conventions and Regional Hamfests
Local clubs in conjunction with the ARRL sponsor annual conventions and hamfests that generally dedicate one or more forums to DXing and contesting. At these events, a person can attend forums on antennas, propagation or other topics; shop for used gear; view new models of transceivers or antennas from vendors or manufacturers; chat with hams from across the state or region; or just hang out. Local hamfests and divisional conventions provide good opportunities for having a vacation, spending time with fellow Amateur Radio enthusiasts, talking with equipment manufacturers, and gaining valuable information from the forums that are offered.

ARRL Divisional Meetings
At least annually, the League's divisional leadership holds general meetings for members within the division. These meetings usually include forums on operating topics such as DXing or contesting. I recall a Mid-Atlantic Division meeting of several years ago, wherein Bernie, W3UR made an excellent presentation via Skype on the top 20 most wanted DX entities and the future plans and possibilities for putting them on the air, just one example of the types of presentations that happen at these meetings.

Buy/Sell/Swap Resources

Where can you turn to for gear to get started or to upgrade your DX station? In addition to visiting your local ham radio suppliers, there are two web sites that you ought to become familiar with.

QTH.com
A ham website created and maintained by Scott Neader, KA9FOX. See **http://www.qth.com**. This site is widely recognized as a resource for buying and selling equipment, and operates off donations made by satisfied users. To access classified ads from the main page click on "Ham Radio Info" at the top center of the page and select "Classified Ads." You are then presented with a list of links to various types of equipment you may wish to browse, or a link for posting advertisements of your own. To search for a particular kind of item, perhaps a transceiver model number, click on "Search" at top left of the page.

eHam.net
eHam is a very extensive Amateur Radio web site, with sections geared towards: the Amateur Radio community, operating and a variety of other topics and resources. See **http://www.eHam.net** At last count, there is a team of about 20 hams taking responsibility for maintaining various sections of this extensive web page.

As far as looking for equipment goes, the site's classified section hosts a large number of ads for equipment of all types, many with pictures showing the condition of the equipment. When trying to decide between different product models (perhaps different transceivers or amplifiers) the "Product Review" section can be helpful; enter a model number or other term, select search, and voila, depending upon how popular an item might be, a long list of reviews will be displayed.

When looking at product reviews, keep in mind they are posted by hams from all walks of life that have varying levels of experience, and perhaps likes and dislikes different from your own. That said, they can still be useful, and perhaps cause you to think of things you may not have previously considered.

Appendix A: Entities of the World

AFRICA

Entity	Prefix	Entity	Prefix
Agalega & St. Brandon Island	3B6; 3B7	Comoros	D6
Mauritius	3B8	Eritrea	E3
Rodriguez Island	3B9	Canary Island	EA8-EH8
Equatorial Guinea	3C	Ceuta & Melilla	EA9-EH9
Annobon Island	3C0	Liberia	EL
Swaziland	3DA	Ethiopia	ET
Tunisia	3V	Mayotte	FH; TO
Guinea	3X	Reunion Island	FR; TO
Bouvet	3Y	Glorioso Island	FT/G; TO
Libya	5A	Juan de Nova, Europa	FT/J; E; TO
Tanzania	5H; 5I	Tromelin Island	FT/T; TO
Nigeria	5N	Crozet Island	FT/W
Madagascar	5R	Kerguelen Island	FT/X
Mauritania	5T	Amsterdam & St. Paul Island	Ft/Z
Niger	5U	Djibouti	J2
Togo	5V	Guinea-Bissau	J5
Uganda	5X	Western Sahara	S0
Kenya	5Y; 5Z	Seychelles	S7
Senegal	6V; 6W	Sao Tome & Principe	S9
Lesotho	7P	Sudan	ST
Malawi	7Q	Egypt	SU
Algeria	7T-7Y	Somalia	T5; 6O
Maldives (spans AS)	8Q	Cameroon	TJ
Ghana	9G	Central Africa	TL
Zambia	9I-9J	Rep. of Congo	TN

Entity	Prefix	Entity	Prefix
Sierra Lione	9L	Gabon	TR
Dem. Rep. of the Congo	9Q – 9T	Chad	TT
Burundi	9U	Cote d'Ivoire	TU
Rwanda	9X	Benin	TY
Botswana	A2	Mali	TZ
The Gambia	C5	Namibia	V5
Mozambique	C8-C9	Heard Island	VK0
Morocco	CN	Chagos Island	VQ9
Madera Island	CT3	Burkina Faso	XT
Angola	D2-D3	Zimbabwe	Z2
Cape Verde	D4	South Sudan	Z8
Ascension Island	ZD8	St. Helena Island	ZD7
Tristan de Cunha & Gough Islands	ZD9	South Africa	ZR-ZU
Prince Edward & Marion Islands	ZS8		

ANTARCTICA

Entity	Prefix	Entity	Prefix
Antarctica	CE9 / KC4	Peter 1 Island	3Y

ASIA

Entity	Prefix	Entity	Prefix
Spratly Island	**	Turkmenistan	EZ
Vietnam	3W; XV	Republic of Korea	HL; 6K-6N
Azerbaijan	4J; 4K	Thailand	HS; E2
Georgia	4L	Saudi Arabia	HZ
Sri Lanka	4S	Japan	JA-JS; 7J-7N
Israel	4X; 4Z	Ogasawara	JD1
Cyprus	5B; C4; P3	Mongolia	JT-JV
Yemen	7O	Jordan	JY
Maldives (spans AF)	8Q	Lebanon	OD
Kuwait	9K	Dem. People's Rep. of Korea	P5

West Malaysia	9M2; 9M4	Bangladesh	S2
Nepal	9N	Turkey	TA-TC
Singapore	9V	Asiatic Russia	UA-UI8,9,0; RA-RZ
Oman	A4	Uzbekistan	UJ-UM
Bhutan	A5	Kazakhstan	Un-UQ
United Arab Emirates	A6	Hong Kong	VR
Qatar	A7	India	VU
Bahrain	A9	Andaman & Nicobar Islands	VU4
Pakistan	AP	Lakshadweep Isl.	VU7
China	B	Cambodia	XU
Scarborough Reef	BS7	Laos	XW
Taiwan	BU-BX	Macao	XX9
Pratas Island	BV9P	Myanmar	XY-XZ
Palestine	E4	Afghanistan	YA; T6
Armenia	EK	Iraq	YI
Iran	EP-EQ	Syria	YK
Kyrgyzstan	EX	UK Soviet Bases on Cyprus	ZC4
Tajikistan	EY		

EUROPE			
Entity	Prefix	Entity	Prefix
Sov. Mil. Order of Malta	1A	Lithuania	LY
Monaco	3A	Bulgaria	LZ
Montenegro	4O	Austria	OE
ITU HQ	4U_ITU	Finland	OF-OI
Croatia	9A	Aland Island	OH0
Malta	9H	Market Reef	OJ0
Andorra	C3	Czech Republic	OK-OL
Portugal	CT	Slovak Republic	OM
Azores	CU	Belgium	ON-OT

Entity	Prefix	Entity	Prefix
Fed. Rep. of Germany	DA-DR	Denmark	OU-OW; OZ
Bosnia-Herzegovina	E7	Faroe Island	OY
Spain	EA-EH	Netherlands	PA-PI
Balearic Island	EA6-EH6	Franz Josef Land	R1/F
Ireland	EI-EJ	Slovenia	S5
Moldova	ER	Sweden	SA-SM; 7S-8S
Estonia	ES	Poland	SN-SR
Belarus	EU-EW	Greece	SV-SZ
France	F	Mount. Athos	SV/A
England	G; GX; M	Dodecanese	SV5; J45
Isle of Man	GD; MD; GT	Crete	SV9; J49
Northern Ireland	GI; MI; GN	San Marino	T7
Jersey	GJ; MJ; GH	Turkey	TA-TC
Scotland	GM; MM; GS	Iceland	TF
Wales	GW; MW; GC	Corsica	TK
Hungary	HA; HG	Kaliningrad	UA2; RA2
Switzerland	HB	European Russia	UA-UI1-7; RA-RZ
Liechtenstein	HB0	Ukraine	UR-UZ; EM-EO
Vatican	HV	Latvia	YL
Italy	I	Romania	YO-YR
Sardinia	IS0; IM0	Serbia	YT-YU
Svalbard	JW	Macedonia	Z3
Jan Mayen	JX	Albania	ZA
Norway	LA-LN	Gibraltar	ZB
Luxembourg	LX		

NORTH AMERICA

Entity	Prefix	Entity	Prefix
United Nations HQ	4U_UN	Navassa Island	KP1
Jamaica	6Y	Virgin Islands	KP2
Barbados	8P	Puerto Rico	KP3-KP4
Bahamas	C6	Desecheo Island	KP5

Cuba	CO; CM	Greenland	OX
Sable Island	CY0	Saba, St. Eustatius	PJ5-PJ6
St. Paul Island	CY9	St. Maarten	PJ7
Guadeloupe	FG; TO	Guatemala	TG; TD
Saint Barthelemy	FJ; TO	Costa Rica	TI; TE
Martinique	FM; TO	Cocos Island	TI9
Clipperton Island	FO; TX	Antigua & Barbuda	V2
St. Pierre & Miquelon Isl.	FP; TO	Belize	V3
Saint Martin	FS; TO	St. Kitts & Nevis	V4
Haiti	HH	Canada	VA-VG; VO; VY
Dominican Republic	HI	Anguilla	VP2E
San Andres & Providencia	HK0	Monserrat	VP2M
Panama	HO-HP	British Virgin Islands	VP2V
Honduras	HQ-HR	Turks & Caicos Isl.	VP5; VQ5
Grenada	J3	Bermuda	VP9
St. Lucia	J6	Mexico	XA-XI
Dominica	J7	Revillagigedo	XA4-XI4
St. Vincent	J8	Nicaragua	YN; H6-7; HT
United States of America	K; W; N; AA-AK	El Salvador	YS; HU
Guantanamo Bay	KG4	Aves Island	YV0
Alaska	KL; AL; NL; WL	Cayman Islands	ZF

OCEANIANA

Entity	Prefix	Entity	Prefix
Figi	3D2	Kingman Reef	KH5K
Conway Reef	3D2	Hawaii	KH6-KH7
Rotuma Island	3D2	Kure Island	KH7K
Timor-Leste	4W	American Samoa	KH8
Samoa	5W	Swains Island	KH8
East Malaysia	9M6; 9M8	Wake Island	KH9
Tonga	A3	Papua new Guinea	P2

Nauru	C2	Tuvalu	T2
Philippines	DU-DZ; 4D-4I	W. Kiribati (Gilbert Isl.)	T30
N. Cook Island	E5	C. Kiribati (Br. Phoenix Isl.)	T31
S. Cook Island	E5	E. Kiribati (Line Isl.)	T32
New Caledonia	FK; TX	Banaba Island (Ocean Island)	T33
Chesterfield Island	FK; TX	Palau	T8
Austral Island	FO; TX	Micronesia	V6
French Polynesia	FO; TX	Marshall Islands	V7
Marquesas Isl.	FO; TX	Brunei Darussalam	V8
Wallis & Futuna Island	FW	Australia	Vk; AX
Solomon Island	H4	Norfolk Island	VK9N
Temotu Province	H40	Willis Island	VK9W
Minami Torishima	JD1	Christmas Island	VK9X
Mariana Islands	KH0	Pitcairn Isl.	VP6
Baker & Howland Island	KH1	Ducie Island	VP6
Guam	KH2	Indonesia	YB-YH
Johnston Island	KH3	Vanuatu	YJ
Midway Island	KH4	Niue	ZK2
Palmyra & Jarvis Island	KH5	Tokelau Isl.	ZK3
Macquarie Island	VK0	New Zealand	ZL-ZM
Cocos (Keeling) Isl.	VK9C	Chatham Island	ZL7
Lord Howe Island	VK9L	Kermadec Island	ZL8
Mellish Reef	VK9M	Aukland & Campbell Islands	ZL9

SOUTH AMERICA

Entity	Prefix	Entity	Prefix
Guyana	8R	Aruba	P4
Trinidad & Tobago	9Y-9Z	Curaco	PJ2
Chile	CA-CE	Bonaire	PJ4

Eastern Island	CE0	Brazil	PP-PY; ZV-ZZ
Juan Fernandez Island	CE0	Fernando de Noronha	PP0F-PY0F
San Felix & San Ambrosio	CE0	St. Peter & St. Paul Rocks	PP0S-PY0S
Bolivia	CP	Trindade & Martim Vaz Islands	PP0T-PY0T
Uruguay	CV-CX	Suriname	PZ
French Guiana	FY	Falkland Island	VP8
Ecuador	HC-HD	South Georgia Isl.	VP8; LU
Galapagos Islands	HC8-HD8	South Orkney Isl.	VP8; LU
Colombia	HJ-HK; 5J-5K	South Sandwich Isl.	VP8; LU
Malpelo Island	HK0	South Shetland Isl.	VP8' LU; CE9; HF0; 4K1
Argentina	LO-LW	Venezuela	YV-YY; 4M
Peru	OA-OC	Paraguay	ZP

Footnote Citations

[i] American Radio Relay League. "The ARRL DXCC List," Newington, CT: April 2012.
[ii] Radio Society of Great Britain. "RSGB IOTA Directory," Bedford, England: 2011.
[iii] "Glen Johnson, W0GJ. K1N Navassa DXpedition. Mission Impossible, or Mission Possible." The Grey Line Report. Vol. 12, No. 1: March 2015.
[iv] "Wirebook V." Press Jones, NU8G. (2009). The Wireman, Inc. Landrum SC. ISBN: 978-0-9726192-1-9.
[v] "Antennas for Everyone." Giles Read, G1MFG. (2010). Radio Society of Great Britain. ISBN: 9781-9050-8659-7.
[vi] ITU Zone Map – EI8IC.
[vii] SK = Silent Key – a dearly departed ham.
[viii] See: http://www.eham.net/reviews/
[ix] See : http://www.arrl.org/files/file/On%20the%20Air/QSL%20Bureau/Outgoing%20QSL%20Bureau%202011.pdf
[x] See: http://www.cdxa.org/w4qsl-bureau/index.php for details.
[xi] See: ClubLog at https://secure.clublog.org/about.php
[xii] "The ARRL DXCC List." April 2012. American Radio Relay League.

Made in the USA
Middletown, DE
18 September 2024

60610578R00068